Deep Learning for Physical Scientists

Deep Learning for Physical Scientists

Accelerating Research with Machine Learning

Edward O. Pyzer-Knapp
IBM Research UK
Data Centric Cognitive Systems
Daresbury Laboratory
Warrington
UK

Matthew Benatan
IBM Research UK
Data Centric Cognitive Systems
Duresbury Laboratory
Warrington
UK

This edition first published 2022
© 2022 John Wiley & Sons Ltd

The right of Edward O. Pyzer-Knapp and Matthew Benatan to be identified as the authors of this work has been asserted in accordance with law.

Registered Offices
John Wiley & Sons, Inc., 111 River Street, Hoboken, NJ 07030, USA
John Wiley & Sons Ltd, The Atrium, Southern Gate, Chichester, West Sussex, PO19 8SQ, UK

Editorial Office
The Atrium, Southern Gate, Chichester, West Sussex, PO19 8SQ, UK

For details of our global editorial offices, customer services, and more information about Wiley products visit us at www.wiley.com.

Wiley also publishes its books in a variety of electronic formats and by print-on-demand. Some content that appears in standard print versions of this book may not be available in other formats.

For general information on our other products and services or for technical support, please contact our Customer Care Department within the United States at (800) 762-2974, outside the United States at (317) 572-3993 or fax (317) 572-4002.

Wiley also publishes its books in a variety of electronic formats. Some content that appears in print may not be available in electronic formats. For more information about Wiley products, visit our web site at ww.wiley.com.

Library of Congress Cataloging-in-Publication Data

Names: Pyzer-Knapp, Edward O., author. | Benatan, Matthew, author.
Title: Deep learning for physical scientists : accelerating research with
    machine learning / Edward O. Pyzer-Knapp, IBM Research UK, Data Centric
    Cognitive Systems, Daresbury Laboratory, Warrington UK, Matthew
    Benatan, IBM Research UK, Data Centric Cognitive Systems, Daresbury
    Laboratory, Warrington UK.
Description: Hoboken, NJ : Wiley, 2022. | Includes index.
Identifiers: LCCN 2021036996 (print) | LCCN 2021036997 (ebook) | ISBN
    9781119408338 (hardback) | ISBN 9781119408321 (adobe pdf) | ISBN
    9781119408352 (epub)
Subjects: LCSH: Physical sciences–Data processing. | Machine learning.
Classification: LCC Q183.9 .P99 2022 (print) | LCC Q183.9 (ebook) | DDC
    500.20285/631–dc23
LC record available at https://lccn.loc.gov/2021036996
LC ebook record available at https://lccn.loc.gov/2021036997

Cover Design: Wiley
Cover Image: © Anatolyi Deryenko/Alamy Stock Photo

Set in 9.5/12.5pt STIXTwoText by Straive, Pondicherry, India

10 9 8 7 6 5 4 3 2 1

Contents

About the Authors

Dr. Edward O. Pyzer-Knapp is the worldwide lead for AI Enriched Modelling and Simulation at IBM Research. Previously, he obtained his PhD from the University of Cambridge using state of the art computational techniques to accelerate materials design then moving to Harvard where he was in charge of the day-to-day running of the Harvard Clean Energy Project – a collaboration with IBM which combined massive distributed computing, quantum-mechanical simulations, and machine-learning to accelerate discovery of the next generation of organic photovoltaic materials. He is also the Visiting Professor of Industrially Applied AI at the University of Liverpool, and the Editor in Chief for Applied AI Letters, a journal with a focus on real-world application and validation of AI.

Dr. Matt Benatan received his PhD in Audio-Visual Speech Processing from the University of Leeds, after which he went on to pursue a career in AI research within industry. His work to date has involved the research and development of AI techniques for a broad variety of domains, from applications in audio processing through to materials discovery. His research interests include Computer Vision, Signal Processing, Bayesian Optimization, and Scalable Bayesian Inference.

Acknowledgements

EPK: This book would not have been possible without the support of my wonderful wife, Imogen.

MB : Thanks to my wife Rebecca and parents Dan & Debby for their continuing support.

1

Prefix – Learning to "Think Deep"

Paradigm shifts in the way we do science occur when the stars align. For this to occur we must have three key ingredients:

1) A fundamental problem, which is impeding progress;
2) A solution to that problem (often theoretical); and crucially
3) The capability to fully realise that solution.

Whilst this may seem obvious, the lack of (3) can have dramatic consequences. Imperfectly realised solutions, especially if coupled with overzealous marketing (or hype) can set back a research field significantly – sometimes resulting in decades in the wilderness.

Machine learning has suffered this fate not once, but twice – entering the so-called AI-winters where only a few brave souls continued to work. The struggle was not in vain, however, and breakthroughs in the theory – especially the rapid improvement and scaling of deep learning – coupled with strides in computational capability have meant that the machine learning age seems to be upon us with a vengeance.

This "era" of AI feels more permanent, as we are finally seeing successful applications of AI in areas where it had previously struggled. Part of this is due to the wide range of tools which are at the fingertips of the everyday researcher, and part of it is the willingness of people to change the way they think about problems so as to use these new capabilities to their full potential.

The aims of this book are twofold:

1) Introduce you to the prevalent techniques in deep learning, so that you can make educated decisions about what the best tools are for the job.
2) Teach you to think in such a way that you can come up with creative solutions to problems using the tools that you learn throughout this book.

1.1 So What Do I Mean by Changing the Way You Think?

Typically in the sciences, particularly the physical sciences, we focus on the model rather than on the task that it is providing. This can lead to siloing of techniques in which (relatively) minor differences become seemingly major differentiating factors.

As a data-driven researcher it becomes much more important to understand broad categorisations and find the right tools for the job, to be skilled at creating analogies between tasks, and to be OK with losing some explainability of the model (this is particularly true with deep learning) at the gain of enabling new capabilities. More than any other area I have been involved with, deep learning is more a philosophy than any single technique; you must learn to "think deep." Will deep learning always be the right approach? Of course not, but with growing experience you will be able to intuit when to use these tools, and also be able to conceive of new ways in which these tools can be applied. As we progress through this journey I will show you particularly innovative applications of deep learning, such as the neural fingerprint, and task you to stretch your imagination through the use of case studies. Who knows, you might even improve upon the state of the art!

Key Features of Thinking Deep

1) *Never get tied into domain specific language, instead use broad terms which describe the process you are trying to achieve.*
2) *Look for analogies between tasks. By forcing yourself to "translate" what is going on in a particular process you often come to understand more deeply what you are trying to achieve.*
3) *Understand the difference between data and information. A few well chosen pieces of data can be more useful than a ton of similar data points.*
4) *Finally, never lose sight of your true goal – at the end of the day this is more likely to be "provide insight into process X" than "develop a super fancy method for predicting X." The fancy method may be necessary, but it is not the goal.*

Thinking back to the three ingredients of a paradigm shift, we remember that one of the major blockers to achieving this is the lack of a capability to implement the solution we have dreamed up. Therefore, throughout this book, I will be teaching you how to use a state of the art deep-learning framework known as TensorFlow and provide real world examples. These examples will not be aimed at squeezing every last piece of performance out of the system, but instead at ensuring that you understand what is going on. Feel free to take these snippets and tune them yourself so that they work well for the problems you are tackling. Finally, I hope that you get as much fun out of coming on this journey with me, as I have had putting it together. I hope that this book inspires you to start breaking down barriers and drive innovation with data not just in your domain, but in everything you do.

2

Setting Up a Python Environment for Deep Learning Projects

2.1 Python Overview

Why use python? There are a lot of programming languages out there – and they all have their plus and minuses. In this book, we have chosen to use Python as our language of choice. Why is this?

First of all, is the ease of understanding. Python is sometimes known as "executable pseudo code," which is a reference to how easy it is to write basic code. Now this is obviously a slight exaggeration (and it is very possible to write illegible code in Python!), but Python does represent a good trade-off between compactness and legibility. There is a philosophy which went into developing Python which states "There should be one (and preferably only one) obvious way to do a task." To give you an illustrative example, here is how you print a string in Python:

```
print("Hello World!")
```

It is clear what is going on! In Java it is a little more obscure, to deal with system dependencies:

```
system.out.println("Hello World!")
```

And in C, it is not obvious at all what is going on (C is a compiled language so it really only needs to tell the compiler what it needs to do):

```
"Hello World!" >> cout
```

In fact, C code can be so hard to read that there is actually a regular competition to write *obfuscated C* code, so unreadable it is impossible to work out what is going on – take a look at https://www.ioccc.org/ and wonder at the ingenuity. So by choosing to use Python in this book even if you are not a regular Python user you should be able to have a good understanding of what is going on.

Deep Learning for Physical Scientists: Accelerating Research with Machine Learning,
First Edition. Edward O. Pyzer-Knapp and Matthew Benatan.
© 2022 John Wiley & Sons Ltd. Published 2022 by John Wiley & Sons Ltd.

Second is the transferability. Python is an interpreted language, and you do not need to compile it into binary in order to run it. This means that whether you run on a Mac, Windows, or Linux machine, so long as you have the required packages installed you do not have to go through any special steps to make the code you write on one machine run on another. I recommend the use of a Python distribution known as Anaconda to take this to a new level, allowing very fast and simple package installation which takes care of package dependencies. Later on, in this chapter, we will step through installing Anaconda and setting up your Python environment.

One other reason for using Python is the strong community, which has resulted in a huge amount of online support for those getting into the language. If you have a problem when writing some code for this book, online resources such as stackoverflow.com are full of people answering questions for people who have had the exact same problem. This community has resulted in the surfacing of common complaints, and the community collectively building solutions to make libraries for solving these problems and to deliver new functionality. The libraries publically available for Python are something quite special, and are one of the major reasons it has become a major player in the data science and machine learning communities.

2.2 Why Use Python for Data Science?

Recently, Python has seen a strong emergence in the data science community, challenging more traditional players such as R and Matlab. Aside from the very intuitive coding style, transferability, and other features described above, there are a number of reasons for this. First amongst these is its strong set of packages aimed at making mathematical analysis easy. In the mid-1990s the Python community strongly supported the development of a package known as numeric whose purpose was to take the strengths of Matlab's mathematical analysis packages and bring them over to the Python ecosystem. Numeric evolved into numpy, which is one of the most heavily used Python packages today. The same approach was taken to build matplotlib – which as the name suggests was built to take the Matlab plotting library over to python. These were bundled with other libraries aimed at scientific applications (such as optimisation) and turned into scipy – Python's premier scientific-orientated package.

Having taken some of the best pieces out of Matlab, the Python community turned its attention to R; the other behemoth language of data science. Key to the functionality of R is its concept of the data frame, and the Python package pandas emerged to challenge in this arena. Pandas' data frame has proven extremely adept for data ingestion and manipulation, especially of time series data, and has now been linked into multiple packages, facilitating an easy end to end data analytics and machine learning experience.

It is in the area of machine learning in which Python has really separated itself from the rest of the pack. Taking a leaf out of R's book, the scikit-learn module was built to mimic the functionality of the R module caret. Scikit-learn offers a plethora of algorithms and data manipulation features which make some of the routine tasks of data science very simple and intuitive. Scikit-learn is a fantastic example of how powerful the pythonic method for creating libraries can be.

2.3 Anaconda Python

2.3.1 Why Use Anaconda?

When you first pick up this book, it may be tempting to run off and download Python to start playing with some examples (your machine may even have Python pre-installed on it). However, this is unlikely to be a good move in the long term. Many core Python libraries are highly interdependent, and can require a good deal of setting up – which can be a skill in itself. Also, the process will differ for different operating systems (Windows installations can be particularly tricky for the uninitiated) and you can easily find yourself spending a good deal of time just installing packages, which is not why you picked up this book in the first place, is it?

Anaconda Python offers an alternative to this. It is a mechanism for one-click (or type) installation of Python packages, including all dependencies. For those of you who do not like the command line at all, it even has a graphical user interface (GUI) for controlling the installation and updates of packages. For the time being, I will not go down that route, but instead will assume that you have a basic understanding of the command line interface.

2.3.2 Downloading and Installing Anaconda Python

Detailed installation instructions are available on the anaconda website (https://conda.io/docs/user-guide/install/index.html). For the rest of this chapter, I will assume that you are using MacOS – if you are not, do not worry; other operating systems are covered on the website as well.

The first step is to download the installer from the Anaconda website (https://www.anaconda.com/download/#macos).

Conda vs. Mini-conda

When you go to the website, you will see that there are two options for Anaconda; Conda; and Mini-conda. Mini-conda is a bare-bones installation of Python, which does not have any packages attached. This can be useful if you are looking to have a very lean installation (for example, you are building a Docker image, or your computer does not have much space for programmes), but for now we will assume that this is not a problem, and use the full Anaconda installation, which has many packages preinstalled.

You can select the Python2 or Python3 version. If you are running a lot of older code, you might want to use the Python2 version, as Python2 and Python3 codes do not always play well together. If you are working from a clean slate, however, I recommend that you use the Python3 installation as this "future proofs" you somewhat against libraries which make the switch, and no longer support Python2 (the inverse is much rarer, now).

So long as you have chosen Anaconda version (not Mini-coda), you can just double click the pkg file, and the installation will commence. Once installation is finished (unless you have specific reasons, accept any defaults during installation) you should be able to run.

```
$ > conda list
```

If the installation is successful, a list of installed packages will be printed to screen.

But I already have Python installed on my computer? Do I need to uninstall?

```
Anaconda can run alongside any other versions of
Python (including any which are installed by the
system). In order to make sure that Anaconda is being
used, you simply have to make sure that the system
knows where it is. This is achieved by editing the
PATH environment variable.
In order to see whether Anaconda is in your path, run
the following command in a Terminal
$> echo $PATH

To check that Anaconda is set to be the default Python
run:
$> which python
```

NB the PATH variable should be set by the Anaconda installer, so there is normally no need to do anything.

From here, installing packages is easy. First, search your package on Anaconda's cloud (https://anaconda.org/), and you will be able to choose your package. For example, scikit-learn's page is at https://anaconda.org/anaconda/scikit-learn. On each page, the command for installing is given. For scikit-learn, it looks like this:

```
$> conda install -c anaconda scikit-learn
```

Here, the –c flag denotes a specific channel for the conda installer to search to locate the package binaries to install. Usefully, this page also shows all the different operating systems which the package has been built from, so you can be sure that the binary has been built for your system.

Task: Install Anaconda, and use the instructions below to ensure that you have TensorFlow installed on your system

2.3.2.1   Installing TensorFlow
2.3.2.1.1   Without GPU   Neural network training can be significantly accelerated through the use of graphical processing units (GPUs), however they are not strictly necessary. When using smaller architectures and/or working with small amounts of data, a typical central processing unit (CPU) will be sufficient. As such, GPU acceleration will not be required for many of the tasks in this book. Installing Tensorflow without CPU involves a single conda install command:

```
$> conda install -c conda-forge tensorflow
```

2.3.2.1.2   With GPU   To make use of TensorFlow's GPU acceleration, you will need to ensure that you have a compute unified device architecture (CUDA)-capable GPU and all of the required drivers installed on your system. More information on setting up your system for GPU support can be found here: https://www.tensorflow.org/install/gpu

If you are using Linux, you can greatly simplify the configuration process by using the TensorFlow Docker image with GPU support: https://www.tensorflow.org/install/docker

Once you have the prerequisites installed, you can install TensorFlow via:

```
$> pip install tensorflow
```

We recommend sticking to conda install commands to ensure package compatibility with your conda environment, however a few earlier examples made use of TensorFlow 1's low-level application programming interface (API) to illustrate lower-level concepts. For compatibility, the earlier low-level API can be used by including the following at the top of your script:

```
import tensorflow.compat.v1 as tf
tf.compat.v1.disable_eager_execution()
```

This has been included in the code examples wherever necessary.

Note: with the latest version of TensorFlow, this command will install Tensor-Flow with both CPU and GPU (if available) support.

2.4 Jupyter Notebooks

Jupyter notebooks provide a method to easily create interactive documents capable of hosting and running Python code. This is a great way to work for many scientific applications, allowing you to incorporate descriptive text alongside executable code – helping others to understand and reproduce your work. In this way, they are an excellent tool for collaboration, and can be used to build living documents in which code can be updated, and visualisations can be easily rerun with new data. To use Jupyter notebooks, you will need to first install Jupyter by running:

```
$> conda install notebook
```

2.4.1 Why Use a Notebook?

Standard Python scripts have their advantages – the code can be contained within a single file that can easily be run and debugged, without the need to step through and run multiple cells of code. Comments can be added to describe blocks of code, and docstrings can be used to provide more comprehensive descriptions of the script's functionality. So why use a notebook?

While Python scripts have a number of advantages while writing code, there are several shortfalls for their use in sharing your work. Comments and docstrings easily become overwhelming when detailed descriptions are required – turning a clean script into an awkward mess of comments and code blocks. This can make it difficult for others to read and understand what's going on. Furthermore, it does not allow you to incorporate plotting or other visualisations alongside your text. This is where notebooks come in.

Notebooks allow you to create a well-structured document that incorporates executable code. The structuring allows you to create a comprehensive, flowing document – in the same way that you would for a paper or report. Unlike a traditional document, notebooks are enriched by their ability to incorporate executable code; code that can be used to run real experiments alongside the text and/or to produce interactive data visualisations. This is particularly important for sharing code and facilitating usability – providing a far more intuitive method of sharing code than a collection of Python scripts.

2.4.2 Starting a Jupyter Notebook Server

To use Jupyter notebooks, you will need to run the Jupyter notebook server. This can be done simply by executing the following:

```
$> jupyter notebook
```

It is recommended that you execute this from the directory you wish to work from, as this makes accessing existing project content and creating new content easy. Once the above command is executed, Jupyter should prompt your browser to open Jupyter's web User interface (UI), taking you to a page that looks like this:

Source: Project Jupyter.

You can then click "new," which will give the option of creating a new file.

Source: Project Jupyter.

To create a new Python notebook, simple click on the Python option (in this case, Python 3). You will then be presented with a new empty notebook.

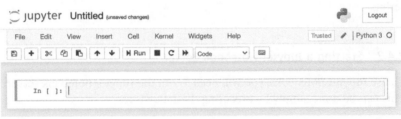

Source: Project Jupyter.

The Jupyter notebook is connected to an iPython kernel, which will allow you to execute Python code in code blocks, such as the block above. Once you have written the code you would like to execute, simply click "run" in the Jupyter UI, or hit shift + enter on your keyboard to execute the cell. Jupyter also has another type of block – *markdown* blocks – which we will explore in the next section.

2.4.3 Adding Markdown to Notebooks

Markdown helps to add structure to documents and improve readability. As with typical markdown, you can easily specify headings and subheadings, as demonstrated in the following example:

```
# Heading 1
## Heading 2
### Heading 3
```

This code produces the following:

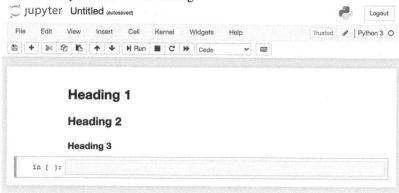

Source: Project Jupyter.

You can also embolden or italicise your text:

```
**Here is an example of emboldening** and
*italicization*.
```

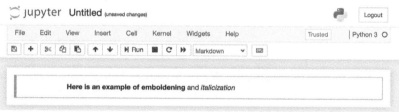

Source: Project Jupyter.

Bullet points or numbered lists can also easily be used:

```
* Here are
* some bullet points
   1. Here is
   2. a list
      1. and sublist
```

Source: Project Jupyter.

Markdown in Jupyter notebooks also allows you to drop in *non-executable* code, in cases where code is being used illustratively, rather than being intended for execution. This is done using backticks:

```
`code = example`
```

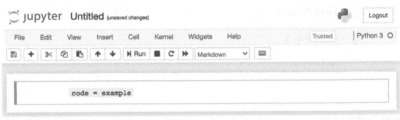

Source: Project Jupyter.

As well as being able to incorporate illustrative code, the markdown also makes it very easy to incorporate mathematical symbols and equations, as it supports LaTeX:

```
$y = \sum_{i=0}^{10}{x^i}$
```

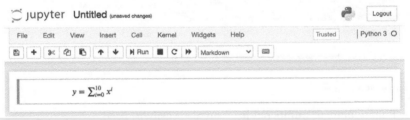

Source: Project Jupyter.

2.4.4 A Simple Plotting Example

Plotting in Jupyter notebooks is an easy and powerful way to visualise data and interact with the visualisations. Below we show a simple plotting example:

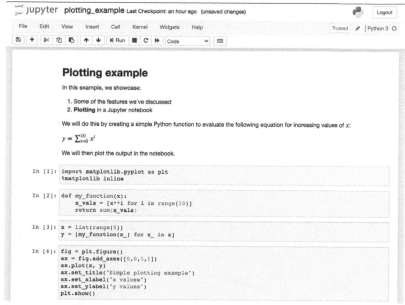

Source: Project Jupyter.

Note the "%matplotlib inline" line in the above code. This will produce an in-line plot within the notebook, as follows:

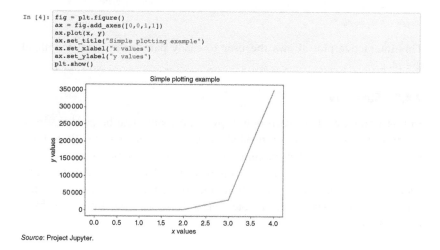

Source: Project Jupyter.

But Jupyter also provides tools for creating interactive plots. This can be achieved by replacing the "%matplotlib inline" instruction to "%matplotlib notebook," producing a plot as follows:

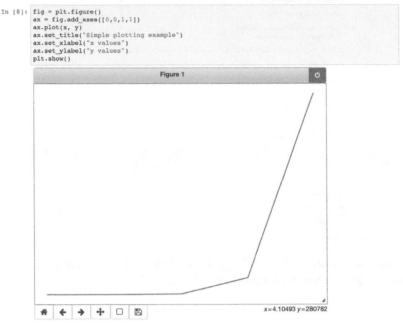

```
In [8]: fig = plt.figure()
        ax = fig.add_axes([0,0,1,1])
        ax.plot(x, y)
        ax.set_title("Simple plotting example")
        ax.set_xlabel("x values")
        ax.set_ylabel("y values")
        plt.show()
```

Source: Project Jupyter.

This interactive plot allows the user to easily pan, zoom, and download the plot.

2.4.5 Summary

In this section, we have seen how Jupyter notebooks can be used to create dynamic documents with embedded code. These notebooks facilitate intuitive knowledge transfer, helping you to convey complex scientific and mathematical concepts with the aid of executable code and interactive visualisations. For more information, and to see further examples of what can be achieved with Jupyter notebooks, please visit the Jupyter website at: jupyter.org

3

Modelling Basics

3.1 Introduction

Before we start our adventure into the world of deep learning, it is a good idea to have a quick refresher on some basics of data-driven modelling. This will include a brief section on how we choose our features (inputs), some best practices such as normalisation of features and targets, and how we tell the model how well it is doing (aka loss functions). If you are experienced at working with data, feel free to skip this chapter; but please do not feel you have to. During my time working in this field, I have often found new insight from hearing someone explain a relatively simple concept in a new or different way. One of the crucial characteristics of a successful machine learning practitioner is the ability to spot analogies. By thinking about techniques from lots of different angles, and explaining them in lots of different ways, we start to learn to abstract the task (job to be done) away from the technique – an ability which is common to many highly successful people in this area.

3.2 Start Where You Mean to Go On – Input Definition and Creation

There is a well-known phrase in data science (and in science in general!) which states that "[if you put] garbage in...[you get] garbage out" – often abbreviated to "GIGO." This is very true in the area of machine learning – there is no magic algorithm which will deliver you amazing results if the relevant information is not contained within the description of the data the model is fed. Indeed, it is often said that a powerful set of features

Deep Learning for Physical Scientists: Accelerating Research with Machine Learning,
First Edition. Edward O. Pyzer-Knapp and Matthew Benatan.
© 2022 John Wiley & Sons Ltd. Published 2022 by John Wiley & Sons Ltd.

coupled to a very simple model are almost always more predictive than a poor set of features coupled to a very sophisticated model. In the past, these features were crafted by hand, to ensure that specific pieces of intuition or knowledge were incorporated into them, although this is changing now (and in Chapter 20 we will see how we can use a convolutional network to learn features directly from raw data).

In the physical sciences, traditional features split generally into two camps – structure-based descriptors and property-based descriptors. Structure-based descriptors describe the physical make-up of a system; for example the position and type of atoms in a molecule, or some parameters which control the size and shape of an aerofoil (such as the National Advisory Committee for Aeronautics (NACA) parameters). These are appealing for two reasons – firstly, typically these descriptors are easy to measure, or cheap to compute – secondly, they should contain enough information to describe many properties of a molecular or physical system; after all, the physics which defines the world is, to a great extent, defined by this structural information.

The second type of feature is based upon properties you know about the input in question. This type of feature is particularly powerful if you believe that the target you are trying to learn depends on some combination of properties you can easily measure. This can be thought of as the machine-learning equivalent of Lipinski's rule of five for drug-like molecules.[1] Another time property-based features are used is when the structure of the system in question cannot be easily measured – for instance if the system is amorphous, or if measuring the structure is very expensive.

3.3 Loss Functions

OK, so we have decided some features which we want to use in our model. The next task is to work out how to tell the model that it is doing well (or badly). A function which does this is known as a loss function. There are many different types of loss function, so picking the right one is essential for the success of your modelling. Since loss functions let us tell the algorithm how to know when it is doing well they really help to frame how you are defining the question which you are asking the model. As you can imagine, they differ between classification and regression. In order to pick the right one, you must first understand whether you are performing a classification or a regression task.

3.3.1 Classification and Regression

In machine learning we typically are performing one of two tasks, either we are trying to find a model which approximates some continuous function (for example trying to develop a model which is able to predict the solubility of a chemical), or one which decides which of the available "buckets" is the most appropriate to place the data point in (for example, deciding whether the chemical is simply "soluble" or "not soluble"). These two tasks are known as regression and classification, respectively.

3.3.2 Regression Loss Functions

3.3.2.1 Mean Absolute Error

The mean absolute error (MAE) is simply the average absolute distance between the predictions and their corresponding ground truths. We can write it down in equation form as follows:

$$\text{MAE}(y, \hat{y}) = \frac{1}{m} \sum_{i=1}^{m} |y_i - \hat{y}_i| \tag{3.1}$$

We can also write it down in code:

```
Import numpy as np

def mean_absolute_error(y, y_hat):
    ```
    Calculate the mean absolute error
    ```
    # First sum the absolute differences
    sum_absolute_error = np.sum(np.abs(y-y_hat))
    # Then divide by the number of entries
    return sum_absolute_error / y.shape[0]
```

The MAE is often used when you suspect that there will be a reasonable number of outliers in the data, as it is more robust to outliers than the more commonly used root mean squared error (RMSE) (up next).

3.3.2.2 Root Mean Squared Error

The RMSE is perhaps the most common loss function in regression. It is the standard deviation of squared differences between the predicted values and the ground truth. You can write down the RMSE as follows:

$$\text{RMSE}(y, \hat{y}) = \sqrt{\frac{1}{m} \sum_{i=1}^{m} (y_i - \hat{y}_i)^2} \tag{3.2}$$

Can you see the similarities between Eqs. (3.1) and (3.2)? Much like the MAE which is based upon the L1 norm (sum of absolutes), the RMSE is based upon the L2 norm (root sum of squares). The higher the order of the norm, the more it weights large differences between prediction and reality, and the more sensitive it is to outliers. Since the RMSE measures standard deviation, we use this when the distribution of errors is broadly bell-shaped, but when outliers change this we can move to less sensitive techniques if it is appropriate.

We can code up RMSE as follows:

```
Import numpy as np

def root_mean_squared_error(y, y_hat):
    ```
    Calculate the root mean squared error
    ```

    #First sum the squared distance
    root_sum_squared_error = np.sqrt(np.sum((a-b)**2))
    # Now divide by the number of entries
    return root_sum_squared_error / y.shape[0]
```

3.3.3 Classification Loss Functions

Things get a little trickier when we start to think about classification of loss functions as simple distance measures do not capture the *question we are trying to ask* the algorithm effectively. Why is this? Well, let us consider the case of pharmaceutical molecules. For any particular problem, there are almost infinitely many more molecules which are not good at treating that problem than there are effective pharmaceuticals (they are called blockbusters for a reason!). Let us say that 1% of molecules are effective (a value of 1) and 99% are not (a value of 0), here is a simple algorithm which guarantees to have high accuracy using the distance measures we have seen up until this point:

```
def not_pharma(molecule):
    return 0
```

As the number of molecules you try grows, the accuracy of this classifier will reach around 99% – which sounds pretty good if you did not know any better!

So why is this? It is due to the small number of positive results in the sample, sometimes called a low base-rate.

3.3.3.1 Precision

One potential way to measure the performance of a classifier is to consider its precision. This metric considers only the accuracy of the positive predictions, that is the true positives (TP), and the false positives (FP). It can be written as:

$$\text{precision} = \frac{\text{TP}}{\text{TP} + \text{FP}} \tag{3.3}$$

Let us consider what that would look like in code, for a binary classifier (0 or 1):

```
import numpy as np

def precision(y, y_hat):
    # First count the true positives (both predict
positive)
    tp = np.sum(np.logical_and(y_hat == 1, y == 1))
    # Now count the false positives (predict positive,
actually negative)
    fp = np.sum(np.logical_and(y_hat==1, y==0))
    precision = tp / (tp+fp)
    return precision
```

However, there is a problem with just considering the precision – it does not take into account the number of positive predictions. One way to "game" this measure is to simply make one perfect prediction; then we would have TP = 1, FP = 0, and perfect precision!

3.3.3.2 Recall

We can remedy this problem with precision by also taking into account the number of false negatives. This is done using a measure called recall:

$$\text{recall} = \frac{\text{TP}}{\text{TP} + \text{FN}} \tag{3.4}$$

We can see this in code for a binary classifier:

```
import numpy as np
def recall(y, y_hat):
    # First count the true positives (both predict
positive)
    tp = np.sum(np.logical_and(y_hat == 1, y==1))
    # Now count the false negatives (predict negative,
actually positive)
    fn = np.sum(np.logical_and(y_hat==0, y==1))
    recall = tp / (tp+fn)
    return recall
```

3.3.3.3 F_1 Score

Recall and precision both measure different, but intertwined, properties of your predictor. By combining them, we can get a single metric embued with the strengths of both. This is known as the F_1 score and is formulated like this:

$$F_1 = \frac{TP}{TP + \frac{FN + FP}{2}} \tag{3.5}$$

This can also be written down as follows:

$$F_1 = 2*\frac{Precision*Recall}{Precision + Recall} \tag{3.6}$$

It can be seen that you can only get a high F_1 score if both the recall and the precision score well. Often, this is a good thing, but it is sometimes not quite what you want. Again, we must think about what the loss function means practically to work out when it is appropriate to use it.

3.3.3.4 Confusion Matrix

The confusion matrix offers a more detailed insight into the performance of a classifier, including all of the potential outcomes. For a binary classifier, it would be formulated as follows:

$$Confusion = \begin{vmatrix} \text{True Neg.} & \text{False Pos.} \\ \text{False Neg.} & \text{True Pos.} \end{vmatrix} \tag{3.7}$$

You can see how previous loss functions, such as precision and recall can be derived from the confusion matrix. In isolation, however, the confusion matrix is more useful as a human-centric description of the performance of a classifier than as a loss function for a learning algorithm. One potential use for it is to use it as a prompt to think about the impact of different (mis-)

classifications in order to rationalise which loss function is most appropriate to use. Nevertheless, let us look at how you might code it up for a binary classifier:

```
def confusion_matrix(y, y_hat):
    confusion_matrix = np.zeros((2,2))
    # Let's calculate the true negatives
    confusion_matrix[0,0] = np.sum(np.logical_and
(y_hat == 0, y==0))
    # Now count the false positives
    confusion_matrix[0,1] = np.sum(np.logical_and
(y_hat == 1, y==0))
    # Now count the false negatives
    confusion_matrix[1,0] = np.sum(np.logical_and
(y_hat == 0, y==1))
    # Now count the true positives
    confusion_matrix[1,1] = np.sum(np.logical_and
(y_hat == 1, y==1))
    return confusion_matrix
```

3.3.3.5 (Area Under) Receiver Operator Curve (AU-ROC)

As we have discussed, unbalanced training sets (for instance when there is a low count of "positive" examples) are both fairly common in scientific problems and can make a nonsense of simple metrics for loss. In binary classification (e.g. does a molecule show pharmaceutical activity or not) there are many approaches to this. One is to use a method such as logistic regression (or a deep network!) which give a probability (0 means no, 1 means yes, 0.5 means no clue) and threshold. For some tasks you might want to threshold towards one extreme, to ensure that the model is very confident in points which are classified as active. Obviously, the result is very strongly correlated with the thresholding value you use, which is less than ideal.

An intuitive way to go about determining the threshold would be to scan through a list of possible values and map the effect it has on the rates of TPs (correctly predicts a positive) and FPs (predicts positive when it should be negative). A good visual way to do this would be to plot them as coordinate pairs, which might give you a plot such as this (Figure 3.1):

If the line tracked along the identity (45° from origin) then we would know that there was absolutely no predictive power in the model at all, whereas the closer it gets to tracking the edge of the upper triangular, the more predictive the model is. This graph is known as the "receiver operator characteristic" curve (commonly shortened to the ROC curve).

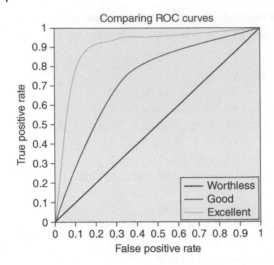

Figure 3.1 Examples of ROC curves.

This is all well and good, but a graph is hardly a useful measure as a loss function for our models – since we want to be able to tune the model automatically to improve it. For this, we need a single number, which we can obtain by integrating under this curve. This is known as the area under the curve – receiver operating characteristic (or AUC-ROC!) and is one of the most common methods for calculating loss in binary classification, since it provides a single number which is independent of the threshold set for classification.

The AUC-ROC can be thought of as the expected FP rate if the ranking is split just after a uniformly drawn random positive. Bearing this in mind, it is possible to draw a rule of thumb for interpreting the quality of a model through AUC-ROC (Table 3.1):

There are a number of ways to calculate the AUC-ROC metric, the one I personally like to use goes as follows. For every pair of predictions (a, b), score a point if:

- a is a positive example and b is a negative example;
- a is greater than b

You can then calculate the AUC-ROC by simply dividing the number of points you score by the total number of pairs you have calculated over. Why would you want to calculate it this way? Well this gives us the opportunity to be a little clever with the code, and speed up the calculation.[2] Instead of naively calculating the score for all pairs, we instead sort the predictions first, and then visit the examples in increasing order of predictions. When

Table 3.1 A rule of thumb guide for understanding AUC-ROC scores.

AUC-ROC score	Interpretation
0.90–1.00	Excellent (A)
0.80–0.90	Good (B)
0.70–0.80	OK (C)
0.60–0.70	Poor (D)
<0.60	Fail (F)

we see a positive example we can then simply add the number of negative examples we have seen so far – a much faster approach. Let us take a look at what this might look like in code form:

```python
import numpy as np


def calc_auc(y_true, y_prob):
    y_true = y_true[np.argsort(y_prob)] #Sort y_true
by the magnitide of the prediction
    count_false = 0
    auc = 0
    n = y_true.shape[0] #How many values are contained
within y_true
    for i in range(n):
        y_i = y_true[i] # Select the ith value of
        count_false += (1 - y_i) # Increases the count
if the true value is negative
        auc += y_i * count_false
    auc /= (count_false * (n - count_false))
    return auc
```

3.3.3.6 Cross Entropy

Imagine that you are sorting through a large amount of data from Hubble, and you want to classify galaxies into one of the four major types: spiral, elliptical, lenticular, or irregular using a set of rules you have implemented on a computer. Since the amount of data to sort through is so large, you want to minimise the number of tests you make on each image. If you have no idea about the underlying distribution of galaxies, the optimal strategy looks like this (Figure 3.2):

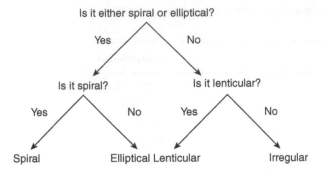

Figure 3.2 Optimal strategy without knowing the distribution.

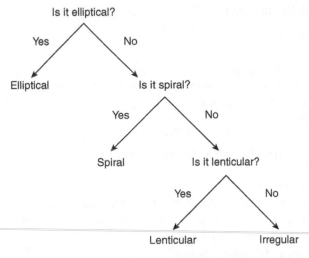

Figure 3.3 Optimal strategy when you know 50% of galaxies are elliptical and 25% of galaxies are spiral.

This is all well and good, but normally we know something about the underlying distribution, and so we can make better decisions. Now imagining that we believe in our data, 50% of the galaxies are elliptical and 25% are spiral. Now, our optimal strategy looks more like this (Figure 3.3):

This is because you are asking the questions in the order of the likelihood of them happening, thus minimising asking an unlikely question. Indeed, the expected number of questions that you have to ask, given an optimal strategy, is known as the entropy. So what has this got to do with the cross-entropy? Well, the cross-entropy is the expected number of questions that you have to ask, given a particular strategy. Clearly, the lower the number of questions, the better the strategy is, and so a low value for the cross-entropy is indicative of a strong strategy.

We can write down the cross-entropy for binary classification as follows:

$$\text{Cross} - \text{Entropy} = -\frac{1}{n}\sum_x y \ln(a) + (1-y)\ln(1-a) \tag{3.8}$$

Although we have demonstrated the cross entropy above using a decision tree, this loss value applies to all classification problems. Remember from our discussion on AUC-ROC that we can write the likelihood (confidence) for the prediction of a "1" as the predicted value, $\hat{y}$, and the likelihood of the prediction of a 0 as $1-\hat{y}$. This means that we can write the total likelihood down as $\hat{y}^y(1-\hat{y})^{(1-y)}$. Why on earth would we want to do this? Well, let us take the log–log of $\hat{y}^y$ is $y\log(\hat{y})$ and log of $(1-\hat{y})^{(1-y)}$ is $(1-y)\ln(1-\hat{y})$. Therefore, to maximise the likelihood, we maximise the following:

$$y \ln(\hat{y}) + (1-y)\ln(1-\hat{y}) \tag{3.9}$$

See any similarities between Eqs. (3.8) and (3.9)? Minimising the cross entropy is the same as maximising the likelihood of the model!

This is not the only thing that cross entropy has going for it. Cross entropy is a cost function which has the ability to avoid one of the major problems of classification cost functions – slow initial learning. This occurs because simple loss functions, such as quadratic losses like RMSE, have the unfortunate property that far away from the "correct" output, the partial derivatives of the cost function (which control the speed of the learning) are very small. This means that the model has a lot of difficulty learning if it strays too far away from a good solution.

It is possible (you can do this as an exercise if you wish) to derive the partial derivatives of Eq. (3.8), and if you do you will get the following:

$$\frac{\delta C}{\delta w_j} = \frac{1}{n}\sum_x x_j(a-y) \tag{3.10}$$

Take a minute to think about this. What Eq. (3.10) is telling us, is that the greater the difference between a and y (the output, and the desired target), the greater the partial derivative of the error with respect to the weights. This means that by using this cost function, our model should learn faster when it is very wrong; thus fixing our *slow learning* problem.

3.4 Overfitting and Underfitting

Once we have built a model, how do we know that it is any good? We could just use the value of whichever loss function we have used to optimise it, but this only tells us how well the model has memorised the answers, not whether it has learned anything about the trends. In order to test this, we will have to test it on examples that the model has not seen before. This is known as model validation. It is important to do this as it starts to distinguish between the model learning transferable relationships between features, and the model simply memorising the data it has been given.

When a model memorises the data, it learns not just the important relationships between features, but also to reproduce the noise contained which is only present because of the potentially arbitrary nature of the construction of the training set. We call this overfitting and it is the bane of many data-driven models, because it is so tempting to give the model as much data as you have and to get the loss value down to as low a number as possible.

One particular cause of overfitting is when you have a complex model, but relatively little data. This is simple to understand – imagine that I have asked you to look at a sequence of numbers and then to predict what the next 10 numbers will be. If that sequence of numbers is small, your brain will provide you an almost infinite number of different rules which will perfectly reproduce the set, but as the length of the sequence (the amount of data) grows, the number of different rules is reduced as more and more evidence steers you towards the correct answer.

Another cause of overfitting is when you have a reasonable amount of data, but your model has a large number of degrees of freedom (in the case of deep learning models this is the number of neurons). Let us think of why this might be with an extreme example using a concept everyone reading this book will be familiar with; linear regression. We are all familiar with fitting a line through X, Y data points. It is of course only possible to fit one perfect line between two X, Y data points. This is because the dimensionality of the plane we are fitting (i), (this is the complexity of the model) is less than the number of data points (ii). However, let us make the model more complex. How many 2D planes can we fit through two data points and still achieve a perfect fit? There are an infinite number!

The opposite of overfitting is a phenomenon known as underfitting. This is when you have not allowed your model to learn a good representation because you have either not allowed it the time or the complexity to do so. An extreme example of this would be to initialise a model with random weights, and then give it one update step to fine tune them. Is it possible that you get an OK model using this method? Sure, but it is unlikely. Another way to underfit a model is to make the task that it is trained to achieve vastly more generic than the task you are interested in applying it to – think of the phrase "jack of all trades master of none."

3.4.1 Bias–Variance Trade-Off

The scale between overfitting and underfitting is known as the bias–variance trade-off. At one end you have models which are focussed on very small areas of problem space, and at the other models which are generalised to perform over a wide range of problem space, potentially at the expense of performance in any one task (Figure 3.4).

We can see how the bias–variance trade-off looks graphically. As you increase the model's complexity, you can reduce the training error substantially, but this often comes at the cost of the validation error – the error on the data which was not used directly to train the machine learning model. Indeed, people often talk about the total error being the sum of the bias, the variance and the irreducible error (sometimes it is simply not possible to have a perfect model). Finding a good solution to our problem is all about finding the region in which we can use model complexity to drive down the error, whilst not increasing it to a point at which it spikes our validation error.

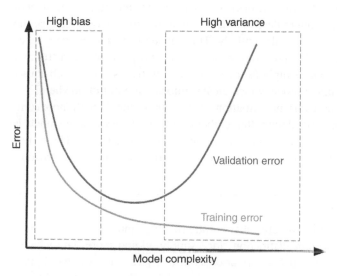

Figure 3.4 A graphical look at the bias–variance trade-off.

Definitions

High bias. Characterised by a high training error. Typically means that that model cannot capture enough predictive structure from the data.

   High variance. Characterised by a high validation error. Typically means that the model has overfit to the training data and does not generalise well.

So now that you can identify high bias or variance situations, is there anything that you can do to deal with the situation when you encounter it? A good start is to think about the reasons why your model might be stuck in a high bias or variance part of the spectrum. Since high bias typically results from models which are not capturing the structure of the data well you can really do two things:

- If the model is amenable (such as a deep learning model) you can train it for a longer time, although, remember that, doing too much of this can push it into a high-variance situation
- If the model is too simple to capture the structure you can increase the complexity.

Training a model for longer is pretty self-explanatory, but how do you increase the complexity of the model? Typically this is achieved by increasing the number of parameters that you are using for fitting (for example the number of neurons in a neural network). Another way is to increase the number of features you are using to describe the inputs to the model, or to change the way you describe the model entirely. This can be particularly effective, as we have discussed that powerful features and a simple model can trump a powerful model with simple features. This is intuitive, since it is obviously impossible to build a model which creates information out of nowhere!

What about high-variance situations? These are arguably both more common, and harder to deal with. Remember that these arise when the model is overfit to the data – it has memorised rather than learned trends. There are a number of ways to "cure" this situation; with the first, and most obvious, being to simply find more data.

A Quick Aside

When I say, find more data, what I really mean is find more *information*. This is an important distinction since simply adding more data which does not contain much information is unlikely to improve the signal to noise problem which is the cause of high-variance models. There are many ways of measuring the information, which will be covered in later chapters, but for now I will assume that the information is distributed evenly across the data which you have access to, and thus adding data is the same as adding information.

Unlike high-bias situations, high-variance situations are often caused by running the learning for too long or having too complex a model. There are a number of ways to reduce the complexity of the model, with the most simple being to reduce the number of features. This can be done by finding redundant features manually and removing them, or through an automated procedure such as principle component analysis (discussed in detail later in this chapter). Another way to reduce the complexity of the model is to restrain the types of solution that it can find. This is known as regularisation (Figure 3.5).

3.5 Regularisation

Regularisation is an attempt to impose constraints into the loss function of a model in order to prevent it from overfitting through explaining the data with an overly complex view of the world.

One way to think about this is with a simple thought experiment. Let us say that we have a model which has two parameters. It fits the data reasonably well, but we are not satisfied, and suspect that it has high-bias. Going through the steps we discussed earlier, we try to add some more features. Unbeknownst to us, however, one of these features is completely unpredictive. Simply by having it, however, we can reduce the training error a little through the increased complexity it introduces. By applying regularisation, we can say that we are happy with a slightly higher training error so long as we reduce the model complexity.

There are lots of ways to introduce regularisation into your loss functions; we will meet a few common ones here, and later in the book tackle some more complex ones aimed specifically at deep learning.

3.5.1 Ridge Regression

The ridge regression method (also known as the L1 regularisation method) aims to reduce the complexity by penalising large values for coefficients. This reduces the complexity of the model as a preference for small coefficients drives towards simple models by reducing the effective number of features. Ridge regression achieves this by adding a penalty term to the model,

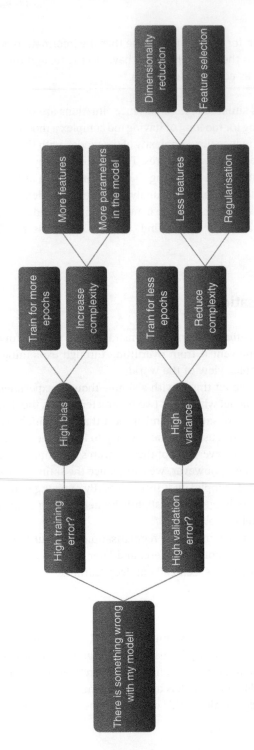

Figure 3.5 A flow chart for dealing with high bias or high-variance situations.

which is determined through summing the squares of the weight terms. We can write down the equation for this penalty term as follows:

$$R_{\text{Ridge}} = \lambda \sum_{j=1}^{p} \beta_j^2 \tag{3.11}$$

where β_j is the value for the jth coefficient in the model, and λ is a tuneable parameter which controls the amount of regularisation. With too large a value for λ, the regularisation will push towards very simple representations, with most features being removed, and with too small a value for λ, the effect of the regularisation will be negligible and we return to the potential for a high-variance model.

Since ridge regression is simply penalising large values for coefficients, it can perform well, even in the presence of highly correlated features. All features will be included in the model (ridge regularisation does not impose sparsity) and the correlation will be reflected in the distribution of the sizes of the coefficients assigned.

3.5.2 LASSO Regularisation

The LASSO regulariser (least absolute shrinkage and selection operator) and operates through adding the absolute value of the magnitude of the weights to the loss function, and is sometimes referred to as L1 regularisation. The LASSO regulariser penalises the model having very large coefficients for weights, and is actually capable of setting the weights of "irrelevant" features to zero, which can be very useful for both complexity limitation and efficiency purposes.

The equation for the LASSO term is very simple:

$$R_{\text{LASSO}} = \lambda \sum_{j=1}^{p} |\beta_j| \tag{3.12}$$

where β_j is the value for the jth coefficient in the model, and λ is a tuneable parameter which controls the amount of regularisation. Again, with a too large a value for λ, the regularisation will push towards very simple representations, with most features being removed, and with too small a value for λ, the effect of the regularisation will be negligible and we return to the potential for a high-variance model.

One of the situations LASSO has an advantage over ridge regression is the case where the inputs have large numbers of features. Since it will sparsify inputs by removing features deemed to be irrelevant, the resulting model is much less computationally challenging, and this can produce significant performance benefits.

One known disadvantage of LASSO regularisation is that it does not perform so well in the presence of highly correlated features. When these are encountered, LASSO will arbitrarily select one of the coefficients and set the rest to zero, which is not ideal if you are exploiting LASSO's ability to do feature selection. Additionally, if you have a small amount of data, LASSO will bound the number of coefficients by the number of samples, which can limit the effectiveness of the model.

3.5.3 Elastic Net

As you can see from the LASSO and ridge regularisation sections, both come with their drawbacks. A very popular solution to these deficiencies is the so-called Elastic Net regulariser. At a high level, Elastic Net is a combination of ridge and LASSO regularisation, and can be written like this:

$$R_{EN} = \lambda_2 \|\beta\|^2 + \lambda_2 \|\beta\|_1 \tag{3.13}$$

It is common to relate λ_2 and λ_1 so that $\lambda_2 = \alpha\lambda$ and $\lambda_1 = (1-\alpha)\lambda$.

It is easy to see that Elastic Net actually captures both LASSO and ridge as special cases. Here the LASSO part of the regularisation generates a sparse model, whilst the ridge part encourages the grouping of variables and stabilises the LASSO regularisation; the best of both worlds!

3.5.4 Bagging and Model Averaging

Bootstrap aggregation, or bagging, involves applying the Bootstrap method to a machine learning model, with the goal of improving the stability of the model. The Bootstrap method comprises the following key steps:

1) From a given population D, sample with replacement M times, each time drawing a sample D_m of n samples
2) For each of the M subsamples, perform some computation, $\theta_m = f(x)$, giving you M values for θ
3) Average over all values of θ to obtain a more accurate estimate for θ

Bagging makes use of the bootstrap method to produce more accurate predictions by combining the outputs of multiple machine learning models. With bagging, we first sub-sample from our population, as in step 1. We then train a machine learning model on each of the M subsamples. During inference, we make a prediction using each of our M models, and average over all M predictions to obtain our final prediction. This method results in greater model stability and reduced overfitting due to the smoothing effect of averaging over models trained on different subsets of the population.

3.6 Evaluating a Model

3.6.1 Holdout Testing

Think back at our discussion on the bias–variance trade-off. One of the things we are trying to avoid is the model that memorises data, thus being unable to distinguish the true trends from the noise which is present in the data set. Of course, these *overfit* models will do extremely well on the data they have been trained upon, which can mislead us as to their power. How to avoid this? The most common way is to split the data into a segment which is used to training, and a *hold out* set which is used for testing the trained model. If the model itself needs tuning (these are called hyperparameters, and we will encounter these later on), we will require a third, validation, set. This is to avoid overfitting the hyperparameters, giving a model which is tuned to just reproduce the test set. Overfitting is everywhere! (Figure 3.6)

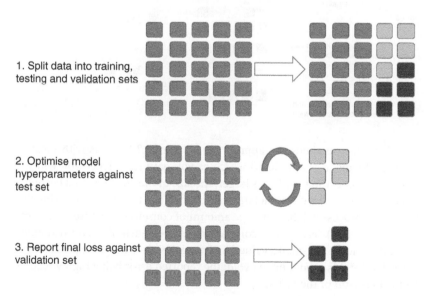

1. Split data into training, testing and validation sets

2. Optimise model hyperparameters against test set

3. Report final loss against validation set

Figure 3.6 Graphical representation of the holdout-validation algorithm.

3.6.2 Cross Validation

A method which splits once, such as a holdout model, does provide some protection against learning the noise in the data, especially if you use a validation set. It is worth thinking about how reliable the validation error is, especially when comparing models, or tuning hyperparameters. It requires, after all, you to completely hold out some data from the model during training – this could be problematic if you do not have much data to begin with, or if your data is not uniformly distributed (in which case it is highly likely, in the case of classification, that your split will be biased towards particular classes).

One way to avoid this, albeit incurring some computational cost at the same time, is to use a technique known as K-folds validation. In K-folds validation, we split the data into K sets (or folds) and cycle through them, each time holding one out and training on the rest.

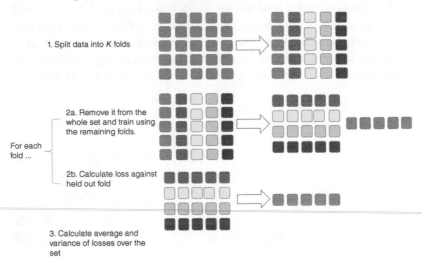

1. Split data into K folds

For each fold ...

2a. Remove it from the whole set and train using the remaining folds.

2b. Calculate loss against held out fold

3. Calculate average and variance of losses over the set

So how do you choose an appropriate value of K? Well, as with most of these problems, it is a trade-off which as you gain experience you will get better at appreciating. As K gets larger, the size of the "testing" fold becomes smaller. Thus, if you choose a large value for K, each training set during the validation process will have a large amount of common structure (and common data points), leading to a correlation between each of the trained models. If you choose too small a value for K, we get back to the same problem we turned to K-folds validation to avoid; the reported error being highly related to the make-up of the test set.

3.7 The Curse of Dimensionality

Working with high-dimensional data (for instance data with a large number of features) has a large number of problems associated with it, most of which are bucketed under the phrase "the curse of dimensionality." This really boils down to three major problems:

1) *High dimensional data is really easy to over-fit.* If you are building a model with a large number of dimensions, and you do not have enough data, you can build a perfect model for this data which does not generalise (remember the linear regression example in the section on overfitting). As the number of dimensions gets large, the amount of data required to build a reasonable model grows exponentially.

2) *Euclidean distances can get less meaningful in high dimensions.* This is because unless the data is varying in all dimensions, the differences in important dimensions can get diluted by the others. This is especially true in very sparse data.

3) *Searching is harder in high dimensions.* Building on from (2), the distance between the centre of the hypercube describing all the dimensions, and its corners is $r\sqrt{d}$ where r is the radius of the largest hypersphere which fits inside the hypercube and d is the number of dimensions. Put another, less abstract, way: the volume of a high dimensional hypercube is made up almost entirely of space around "corners" and almost no "middle." This can create many problems when searching over spaces where the number of potential dimensions is high (e.g. for searching for hyperparameters for complex models).

Does this mean all is lost? Not quite. Whilst all of these concerns are true, and will affect you in your work building models from data, they also are based on the assumption that all your dimensions are independent and identically distributed (the iid assumption). As soon as there is some correlation in your data, this structure can be exploited to improve the situation. For example, you may be able to remove unnecessary dimensions with a dimensionality reduction technique, or build a distance measure which is less sensitive to these problems.

3.7.1 Normalising Inputs and Targets

As scientists we are very comfortable with the fact that data, and features, can exist on many different scales. It is much harder for algorithms to understand this, though. One way in which this is addressed is through the normalisation of features, which results in feature descriptors which all exist within a common range. One advantage of this is that the model does not have to spend time learning how to scale the inputs (which may not even be possible), and another is the effect normalisation has on the learning procedure itself.

Different scales in feature space will translate themselves into different scales in parameter space – since any model will have to learn how to combine these features together. This can lead to a complex loss function which can be hard to optimise. This can be understood fairly simply. When optimising, we want the loss function to look as concave and simple as possible. An ideal optimisation problem is shown in Figure 3.7b, it is clear to see how to efficiently move towards the centre (minimum). If the values for each feature exist on different scales, the loss topology gets warped. Figure 3.7a shows how simply changing the scale of one parameter can have significant effects on the loss surface, making the optimisation harder. Now remember that neural networks can have billions of parameters – it is clear to see how different scales of features can make your life very difficult.

It is typical to normalise your features and targets to exist with a zero mean and unit variance, a technique known as the standard scalar approach. You may, however, want to normalise your data differently based upon what you know about the inputs. For example, in image processing you may simply divide the red green blue (RGB) value of each pixel by 255 to ensure that the value exists in the range 0,1.

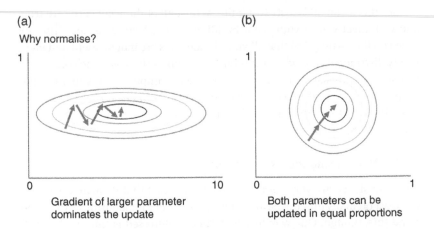

Figure 3.7 The effects of different scales on a simple loss function topology.

3.8 Summary

In this section, you have learned how to do the basics of data processing,

- You understand the difference between classification and regression and when to choose one over the other
- You understand how to identify model overfitting and underfitting – and what to do when you encounter it
- You understand model complexity, and how to introduce regularisation into your models (and what the consequences are)
- You understand loss functions, and how to select the appropriate function for your problem
- You understand the curse of dimensionality, and how to address high-dimensional data
- You understand why and when to normalise your data

Now it is time to put some of this knowledge into practice!

Notes

1 Lipinski's rule of five is a rule of thumb used in drug discovery to determine how drug-like a chemical compound is. It states that a drug-like molecule can have no more than five hydrogen bond donors, no more than 10 hydrogen bond acceptors, a molecular mass of under 500 Da, and a $logP$ of less than 5.
2 I was first shown this approach by my colleague Jean-Francois Puget.

4

Feedforward Networks and Multilayered Perceptrons

4.1 Introduction

Although it might seem unlikely, artificial neural networks have been around for quite some time. The neurons, in an artificial neural network are based on a computational model of how biological synapses work, although the "learning" mechanisms are less biologically inspired.

In this section we will build up the understanding of the most basic form of deep learning system, the multilayer perceptron (MLP). Many of the tricks we learn here can be applied to all the different types of deep networks we will meet in the following chapters of this book.

4.2 The Single Perceptron

4.2.1 Training a Perceptron

Whilst the neural network may at first seem very complex, we can understand the training process by first considering a single neuron. This kind of model is sometimes known as a perceptron. The algorithm for training perceptrons is very simple, and goes as follows:

1) Initialise the weights (Figure 4.1)
2) Calculate the output of the neuron
3) Compare with the desired output, calculate error
4) Update the weights based upon the gradient of the error
5) GOTO step 2

Deep Learning for Physical Scientists: Accelerating Research with Machine Learning,
First Edition. Edward O. Pyzer-Knapp and Matthew Benatan.
© 2022 John Wiley & Sons Ltd. Published 2022 by John Wiley & Sons Ltd.

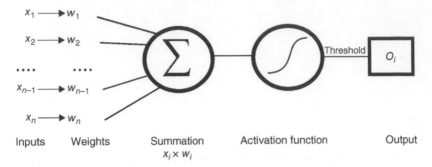

Figure 4.1 An overview of a single perceptron learning.

Each cycle is known as an *epoch*. There are some key components of a perceptron that we should understand, to fully appreciate what goes into training a more complex neural network, such as a multi-layered perceptron. These control how signals pass through neurons (activation functions) and how this output is translated into learning (back-propagation), and we will delve into them in more detail now.

4.2.2 Activation Functions

When a signal is passed to a neuron, it must decide whether to fire or not. This is done using an activation function. Biological neurons have a binary activation function, which looks like a step function. In artificial neurons, however, this has proven to be fairly ineffective for building high performance networks, and so is pretty much never used. The activation functions which are used instead fall into two main types, saturating and non-saturating.

The simplest activation functions are the so-called sigmoidal functions, such as tan*h* and the logistic function.

The tan*h* function is defined as follows:

$$\varphi(\alpha) = \tan h(\alpha) \tag{4.1}$$

Another popular activation function is the logistic function, which looks like this (Figure 4.2):

$$\varphi(\alpha) = \frac{1}{1 + e^{-\alpha}} \tag{4.2}$$

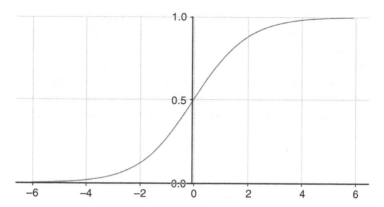

Figure 4.2 The logistic function.

Let us put that down in code:

```python
def logistic(features):
    """
    Compute the logistic activation function. Assumes
    that the features are a tensorflow tensor of floats.
    """

    signal = 1. / (1. + np.exp(-1.*features)

    return signal

def tanh(features):
    """
    Compute the tanh activation function. Assumes that
    the features are a tensorflow tensor of floats.
    """

    signal = np.tanh(features)

    return signal
```

4.2.3 Back Propagation

So you can take an input and send the signal through a perceptron to give an output, but how do you translate this into learning? This is known as "training the network." Back propagation is the most popular method for training neural networks. It works by taking the loss function calculated at the end of an epoch and distributing it back through the network. This "distributed" error is used to adjust the weights. Think of it this way – if the error is large, the weight(s) are clearly a long way off, and so they are adjusted by a large amount. If, however, the error is small then the weight(s) are pretty close, so we should only nudge them a little to improve the network.

Before we think about a deep neural network, we can work through on a simpler case, the single neuron.

For a single neuron, you can calculate the instantaneous error function (loss function for an epoch) $e(n)$ for the output of the neuron:

$$e(n) = y_{\text{label}} - y_{\text{out}} \tag{4.3}$$

One common way of expressing this error is the quadratic cost function, which can be written as:

$$\varepsilon(n) = \frac{1}{2}e^2(n) \tag{4.4}$$

We can then use the gradient of this to update the weights:

$$\Delta w(n) = \frac{\delta \varepsilon(n)}{\delta w(n)} \tag{4.5}$$

After we have updated the weights we start the whole process again and repeat until we hit some predetermined termination condition (the most simple of these being a fixed number of repeats).

So why does this work? Let us go back for a second and think about how an activation function looks like. We have already met some of the simplest ones, the sigmoidal activation functions tan*h* and logistic. Extreme values in this curve result in a strong 1 or 0 signal – that is to say that the neuron is sure of its output. A "middling" value (say 0.4) indicates that the network is unsure about its output; but *how* unsure? We can get a measure of this through the gradient of the error function (Figure 4.3).

If we look carefully at the weight update rule (above) we can see that we can rewrite it in terms of a local gradient, $\delta_j(n)$

$$\delta_j(n) = e_j(n)\varphi'\big(v_j(n)\big) \tag{4.6}$$

Where, the local gradient is the product of the error signal of the neuron and the derivative of the activation function $\varphi'(v_j(n))$. Figure 4.4 shows how the derivatives of the activation function vary with the input. We can see that when the value is extreme, the derivative is small, but in the middle of the

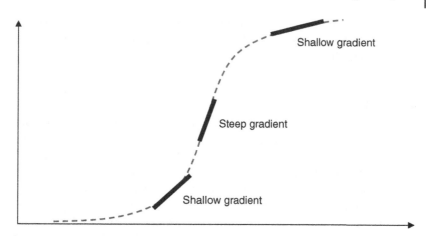

Figure 4.3 Derivatives of the logistic function.

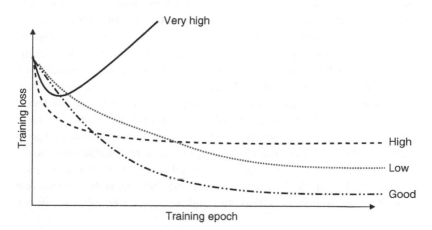

Figure 4.4 How learning rate can affect the training, and therefore performance, of a network.

curve where the neuron is most unsure, the derivative is largest. This means that when we multiply the error signal by the gradient, the update is large when the neuron is unsure, but small when the network is highly confident.

4.2.3.1 Weight Initialisation

Obviously, the weights must start with some value, and this is known as the initialisation. Classically, the weights of the network are initialised through randomly sampling from a normal distribution with a mean of 0 and a standard deviation of 1, but we will take a look at more fancy initialisation schemes later in the chapter.

4.2.3.2 Learning Rate

The learning rate determines how much the weights of the network are modified with respect to the gradient. If we have a low value for the learning rate, the weights are modified only slightly, but if we have a high value the weights are modified more significantly. This can be interpreted as a step size – a higher learning rate corresponds to larger steps, leading us to progress more quickly through the loss manifold, while a lower learning rate corresponds to smaller steps. At first, progressing quickly may sound like a good idea – surely, progressing more quickly gets us to the end goal – a trained network – sooner? Unfortunately things are not quite that simple. Due to the shape of the loss manifold, having a high learning rate is often detrimental. This is as we will tend to step over important minima, and will end up in a suboptimal plateau, as demonstrated in Figure 4.4. This effect is further compounded with a very high learning rate – as we will skip over key regions in the loss space entirely, potentially resulting in significantly worse performance than we had with our random initialisation. So, surely this means a low learning rate must be optimal? Again, the truth is a little more nuanced. While a low learning rate should eventually result in a low loss, and therefore good values for our weights, it will also mean that our network will train very slowly. The truth lies somewhere between these two extremes – a good learning rate will allow training to progress at a reasonable pace, while ensuring that we do not step over crucial minima in our loss space (Figure 4.4). In fact, ideally we want something that adapts the learning rate depending on how well we're doing, and we will cover this later in the chapter.

4.2.4 Key Assumptions

- The cost function can be written as an average over the individual costs calculated for data points
- The cost can be written as a function of the output activations (i.e. it is not dependent on training parameters such as the learning rate or bias)

4.2.5 Putting It All Together in TensorFlow

Now we will see how to put this together using TensorFlow. In order to really look under the hood, we will do this example using low-level application programming interfaces (APIs), which require a few extra headers.

```python
import tensorflow.compat.v1 as tf
import numpy as np
tf.compat.v1.disable_eager_execution()


class SingleNeuronRegression:
  def __init__(self, X_train, y_train, session):
    """
    This is the area for setting up our TF variables
and initializing the network
    :param X_train: an array of features
    :param y_train: an array of labels (or targets)
    :param session: a TF session
    """
    self.session = session #This allows us to keep a
single session open
    self.X_train = X_train
    self.y_train = y_train
    self.features = tf.placeholder(tf.float64, [None,
X_train.shape[1]]) # This is our TF placeholder for
arrays that look like features
    self.labels = tf.placeholder(tf.float64, [None,
y_train.shape[1]]) # This is our TF placeholder for
arrays that looks like labels
    self.initialize_weights() # This randomly
initializes our weights
    self.init = tf.global_variables_initializer()

  def initialize_weights(self):
    """
    This function randomly initializes weights using
sampling from a normal distribution
    :return:
    """
    random_weights = np.random.normal(size=X_train.
shape[1]) # Draw from a normal distribution of mean 0,
std 1
```

```
    self.weights = tf.Variable(np.array([[i] for i in
random_weights]), tf.float64) # Place these weights
into a TF variable

  def train(self,epochs=100):
    """
    Here we train the neuron using simple gradient
descent and a l2 loss function
    :param epochs: The number of epochs to run for
training the neuron
    :return:
    """


    self.session.run(self.init) # Open the TF session
    self.output = tf.nn.sigmoid(tf.matmul(self.
features, self.weights)) # Calculating the output of
the neuron
    self.loss = tf.reduce_mean((self.labels - self.
output)**2) # Calculating the loss of the prediction
    self.optimizer = tf.train.GradientDescentOptimizer
(1) # Set the algo for optimizing the weights with
gradient descent
    # 1 means that learning rate is not contributing,
we will learn more about this later
    training = self.optimizer.minimize(self.loss)
    for i in range(epochs):
        self.session.run(training, feed_dict={self.
features:self.X_train, self.labels:self.y_train}) #
Run an epoch in the session
        error = self.session.run(self.loss, feed_dict=
{self.features:X_train, self.labels:self.y_train}) #
Calculate the loss for monitoring purposes
        print('Training Cost: {}'.format(error))
    return error
  def predict(self, X_test):
    """
    Here we use the trained network to predict on some
new features
    :param X_test: an array of features to predict on
```

```
:return: Predicted target values for X_test
    """
    output = self.session.run(self.output, feed_dict=
{self.features:X_test}) # Run X_test through the
network
    return output
```

4.3   Moving to a Deep Network

As you can see, a single perceptron is a flexible model, but has some pretty severe limitations with regards to the complexity of the models it can approximate.

Early neural network models were simple networks of these perceptrons consisting of one or two layers. This limited the network to learning relatively simple combinations of input features, which severely limits their effectiveness as there is no opportunity to learn complex interdependencies.

The natural way to extend this is to create a network of neurons, and start to move into deeper models. The simplest construction of a deep neural network is known as a multilayered perceptron. When we move into MLPs, we discover a new kind of layer – the hidden layer. Hidden layers, as their name suggests, are hidden from the outside world. That is to say, they are sandwiched between the input and the output layers. In the most common cases, every neuron in a layer is connected to every neuron in the next layer – the so-called *fully connected* feedforward neural network (Figure 4.5).

Why is this important? Well with the introduction of one or more hidden layers we now allow the neural network to be able to model much more complex relationships. Why is this? Let us think of the simple three layer network – one input, one hidden, and one output layer. Without the hidden layer, the signal passed to the output layer must be a combination of inputs, or features. The hidden layer can be thought of as learning a new representation which describes features of the features, which are refined during the learning process. The hidden layer is the first small step away from the so-called "statistical" machine learning methods, and towards more complex learning machines.

As you can imagine, the introduction of a hidden layer (or many hidden layers!) can introduce some complications to the maths. Back propagation still works, but now we have two situations, the neuron is either one of the input or output neurons, or it is one of these new, hidden, layers. When we are propagating the error back through the network, we can now hit two

Input layer   Hidden layer   Output layer

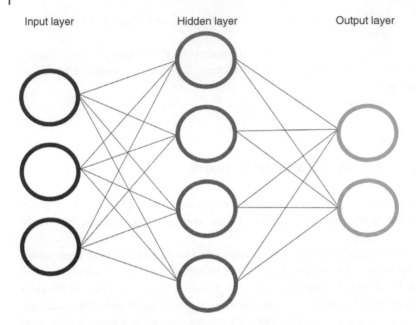

Figure 4.5 A schematic of a multilayer perceptron.

different situations. If a neuron is in the output layer, we know the desired response, so it is easy to compute the local gradient, just like before (recall $\delta(n) = e_j(n)\varphi'(v_j(n))$). If the neuron is in the hidden layer, however things are a bit more tricky as it is not clear how to calculate the error when we do not know what the signal of a hidden layer *should* be. One way to achieve this, is to recursively determine the error signal and work backwards

$$\delta_j(n) = \varphi'(v_j(n))\sum_k \delta_k(n)w_{kj}(n) \tag{4.7}$$

In this case, the local gradient is the product of the derivative $\varphi'(v_j(n))$ and the weighted sum of the gradients associated with neurons in the hidden or output layers to which it is connected.

```
class MultiLayerPerceptronRegressor(object):
    """

    A class for building an MLP for regression tasks
using TensorFlow
    """
```

```python
    def __init__(self, X, y, layers, std_init=0.1):
        """

        Initialization
        :param X: Training features
        :param y: Training targets
        :param layers: list of layer sizes
        :param dropout: list of dropout probs
        :param activation: activation function for network
        :param std_init: standard deviation for the random
init of weights
        """
        self.nlayers = len(layers)
        self.layer_description = layers
        self.layers = None
        self.weights = None
        self.biases = None
        self.X = X
        self.y = y
        self.std_init = std_init

    def activation(self):
            return np.tanh

    def _random_init_weights(self):
        """

        Randomly initialize the weight matrix
        :return:
        """
        weights = []
        # Input layer
        weights.append(tf.Variable(tf.random_normal
([self.X.shape[1], self.layer_description[0]],
stddev=self.std_init)))
        # Hidden layer_description
        for l in range(1, self.nlayers):
            weights.append(tf.Variable(tf.random_normal
([self.layer_description[l - 1], self.
layer_description[l]], stddev=self.std_init)))
        # Output layer
        weights.append(tf.Variable(tf.random_normal
([self.layer_description[-1], self.y.shape[1]],
```

```
stddev=self.std_init)))
    self.weights = weights
  def _eval_train(self, _X):
    """
    Builds the TF layers
    :return:
    """
    layers = []
    # Input layer
    layers.append(self.activation(tf.add(tf.matmul
(_X, self.weights[0]))))
    for l in range(self.nlayers):
      layers.append(self.activation(tf.add(tf.matmul
(layers[l], self.weights[l+1]))))

    # Output layer
   out = self.activation(tf.add(tf.matmul(layers[-1],
self.weights[-1])))
    return out


  def train(self, cost=tf.contrib.losses.
mean_squared_error, optimizer=tf.train.SGDOptimizer
(learning_rate=1e-3)
        , epochs=1000, batchsize=100):
    """
    Train the neural network
    :param cost: tf loss function
    :param optimizer: tf optimizer
    :return:
    """
    init_all = tf.initialize_all_variables()
    pred  = self._eval_train()
    correct_prediction = tf.equal(tf.argmax(pred, 1),
tf.argmax(self.y, 1))
    accuracy = tf.reduce_mean(tf.cast
(correct_prediction, tf.float32))
    Xph = tf.placeholder(tf.float32, [None, self.X.
shape[1]])
    yph = tf.placeholder(tf.float32, [None, self.y.
shape[1]])
    cost = tf.reduce_mean(tf.nn.l2_loss(pred, self.y))
```

```
optimizer = optimizer.minimize(cost)
nbatches = int(self.X.shape[0] / batchsize)
with tf.Session as sess:
   for epoch in range (epochs):
                    sess.run(optimizer, feed_dict={Xph:
batch_xs, yph: batch_ys})
                        train_acc = sess.run(accuracy,
feed_dict={X: self.X, y: self.y})
```

4.4 Vanishing Gradients and Other "Deep" Problems

As our network gets deeper and deeper, we have some new problems to think about. One of the most important is the problem of vanishing gradients. Remember how we propagate the error signal back through the network in back propagation? A signal from an error is spread over the final hidden layer, which is then spread over the penultimate hidden layer (etc. etc.). It is easy to see that as the network gets deeper, the error signal (and thus its gradients) get diluted. This could mean that there is very little signal left by the time the back propagation algorithm reaches the early layers, which means that it is very hard for them to learn. This can really mess with the learning ability of the whole network and kill the usefulness of the system. This is at conflict with what we have seen about the ability of deeper networks to learn more complex relationships, and is fairly intuitive when we think about how people learn. It is easy to learn things, where we can easily understand how to combine the features to get the outputs, but when these combinations get complex, it is much harder for us to learn.

Another related problem is the saturation of the activation function. This happens when the activation function receives a signal which is either very large or very small. A simple inspection of the form of the so-called saturating activation functions $tanh$ and logistic will show you that these extreme signals will result in either a 1 or a 0. What is most important is what this will do to the gradient; since this is what we use to back propagate the error. With extreme values, the gradient of the activation function is essentially 0, which means that the learning is stalled as no information can be obtained from the back-propagating gradients. As a rule of thumb, the logistic function suffers more from this problem, as the mean is 0.5, not 0, like $tanh$.

- A small signal will vanish until it is too small to be useful
- A large signal will grow until it is too large to be useful

Clearly, we need to address this to build the sophisticated deep learning systems which we desire. There are two methods for combating this: the first is to use gradient clipping, and the second is to use one of the family of the so-called non-saturating activation functions.

4.4.1 Gradient Clipping

Gradient clipping ensures that the maximum and minimum values of the gradient lie within a specified range. This is done by applying upper and lower thresholds to the gradient value, often through the method of L2 Norm Clipping. Using this method, the L2 norm of the gradients is computed, and if this value falls above a maximum value the gradient value will be modified so that its norm is equal to the maximum specified value. For example, we could specify that the L2 norm should never exceed 0.5. In this case, if the norm of the gradient exceeds 0.5, the gradient value will be modified so that the norm is equal to 0.5. In Keras, this can be achieved using the "clipnorm" argument when defining your optimiser:

```
opt = SGD(lr=0.01, clipnorm=0.5)
```

4.4.2 Non-saturating Activation Functions

4.4.2.1 ReLU

The most common non-saturating activation function is the rectified linear unit (ReLU). This can be written as (Figure 4.6):

$$A_{\text{ReLU}}(x) = \max(0, x) \tag{4.8}$$

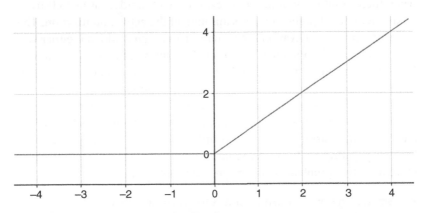

Figure 4.6 Plot of ReLU activation function.

Due to the fact that this activation cannot saturate (since there is no maximum value), the gradients do not disappear as they do in a tan$h$ or logistic activation. Another side effect of the use of the ReLU activation function is a small (but sometimes significant) speedup in the inference time of the trained network; this is because the ReLU activation function is faster to compute than a logistic or tan$h$. This can also be noticed by a slightly faster time to compute an epoch during training. It should be noted, however, that speed in deployment is rarely the make or break for a machine learning solution, especially in the scientific arena, and it is unwise to trade this off too heavily against accuracy.

The ReLU activation has been used to much effect, and is now the most commonly used activation function in deep learning applications, particularly in computer vision. This does not mean, however, that it is without fault. Some problems with the ReLU activation are:

- Not differentiable at 0 – can cause problems with gradient descent (however it is differentiable everywhere else so this rarely causes problems)
- Not 0 centred – a positive mean can pass a biased signal into the following layers and slow down learning
- Unbounded – the value can get larger infinitely.

The largest problem with ReLU, however, is known as the dying neuron problem. In some situations, neurons with ReLU activations can find themselves in a situation where they never fire, whatever the input. This derives from the fact that ReLU will always output zero for a negative input signal, so if all the input signals are negative, then the ReLU becomes unable to discriminate between them. One of the more common reasons this might happen is if you make a particularly bad step whilst performing gradient descent; often the fault of too high a learning rate.

This obviously kills the gradient, thus stalling the learning. Even worse, it is not uncommon for a large number of neurons in a network to die during training. This results in the effective size of the network being much smaller than you might expect it to be, which limits the learning capacity of the remaining neurons, negatively affecting performance.

Let us look at this in code:

```
def relu(features):
    """
    Compute the ReLU activation function. Assumes that
the features are a
    tensorflow tensor of floats.
```

```
"""
signal = math_ops.maximum(0, features)
return signal
```

4.4.2.2 Leaky ReLU

We can address the dying ReLU problem by adjusting the activation function to have a small "leak" to stop the activation getting stuck. This is defined as a non-zero gradient below 0.

$$A_{ReLU-L}(x) = \max(0, x) \text{ if } x > 0 \text{ or } \min(-ax, 0) \text{ if } x < 0 \tag{4.9}$$

Which looks like this (Figure 4.7):

The exact slope (a) can be either set by the user at the start of training, or learnt dynamically during back propagation, although this second option has an accompanying risk of overfitting if the dataset is small, or the information contained within it is particularly narrow.

Let us look at this in code:

```
def leaky_relu(features, alpha=0.2):
    """
    Compute the Leaky ReLU activation function.Assumes
    that the features are a
    tensorflow tensor of floats.
    """
    signal = math_ops.maximum(alpha * features,
features)
    return signal
```

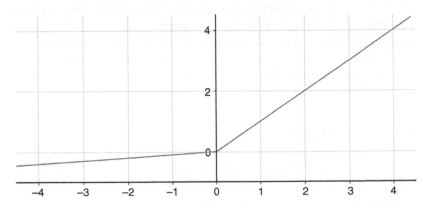

Figure 4.7 Plot of leaky ReLU activation function.

4.4.2.3 ELU

Another way of countering the dying neuron problem is to use exponential linear units (ELU), which can be expressed as:

$$A_{ELU} = \max(0, x) \text{ if } x > 0 \text{ or } \alpha(\exp(x) - 1) \text{ when } x < 0 \qquad (4.10)$$

Which looks like this (Figure 4.8):

Whilst this function is costlier to evaluate, it often shows much better results. This is because the modified behaviour when $z < 0$ pushes the mean closer to 0, which can help the learning process progress smoothly.

4.4.3 More Complex Initialisation Schemes

Another way to control the gradient problem is to use a more sophisticated weight initialisation scheme. Due to the sophisticated architecture of deep networks, and the complex problems to which they are applied, weight initialisation often has a significant impact on network convergence. Depending on the values of the initial weights, the network could be relatively close to convergence, far from convergence, or in a state for which convergence is not even possible. This impact on network convergence is entirely due to the values of the gradients and the resulting trajectory along the loss function given the initial weights, so better initialisation schemes often result in better models, and in some cases are crucial. Here, we discuss two initialisation schemes – the first proposed by Xavier Glorot and Yoshua Bengio, and the second proposed by Kaiming He et al.

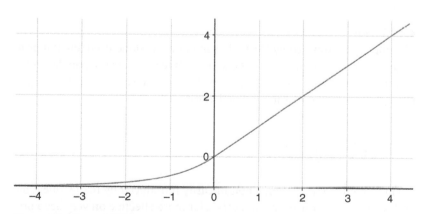

Figure 4.8 Plot of ELU activation function.

4.4.3.1 Xavier

The Xavier initialisation scheme looks to ensure that the initial weight values are not too big or too small, and that they fulfil the symmetry breaking requirements of weight initialisation. This helps to reduce the likelihood of exploding or vanishing gradients, while also ensuring that initial weight values are sufficiently diverse for training to proceed effectively. To achieve this, the method assigns weights from a Gaussian distribution with zero mean and a given variance. But what is a suitable value for the variance? As Glorot and Bengio point out in their paper, it is beneficial to have the variance of the input gradient equal to that of the output gradient. To achieve this, Glorot and Bengio average over the number of inputs and outputs when setting their variance, and the weights are sampled from the following distribution:

$$N\left(0, \sigma = \sqrt{\frac{2}{n_{\text{inputs}} + n_{\text{outputs}}}}\right)$$

It is worth noting that the variance of outputs of each layer need to be the same as variance of inputs (for logistic)

$$\text{Initialise as } N\left(0, \sigma = \sqrt{\frac{2}{n_{\text{inputs}} + n_{\text{outputs}}}}\right)$$

$$\text{or Uniform } (-r, +r), r = \sqrt{\frac{6}{n_{\text{inputs}} + n_{\text{outputs}}}}$$

4.4.3.2 He

The approach proposed by He et al. is built on the same fundamental principles as Xavier initialisation, but extends this to work for the popular ReLU activation functions. For this approach, the variance only considers n_{inputs}, not n_{outputs}, and the distribution is defined as:

$$N\left(0, \sigma = \sqrt{\frac{2}{n_{\text{inputs}}}}\right)$$

This accounts for the ReLU non-linearity introduced by the rectification, and work has demonstrated that this is far more effective on very deep problems when compared to.

4.4.4 Mini Batching

One of the ways we can improve our learning is to use mini-batching. The concept for mini-batching is simple; you split your data into batches and then you pass each batch through the network, propagating the error on a per-batch basis. Why would you want to do this though? Let us think about the information contained within a data set. Unless the information has been deliberately and efficiently collected, it is likely that clusters exist within the data. When you pass this data through the network you are passing many data points which are almost the same to the network, which means the information which is gained by the network seeing some points is less than seeing other, rarer, points. When you split up the data into batches you are making many more, smaller, passes through the network where the contributions of each data point to *that pass through the network* are much more even. Due to this redundancy in the data you are using to make the mini-batches, the update that you make based upon one mini-batch will be approximately the same as the one you would have made had you used the whole set, but at a fraction of the cost. Thus, by the time you have passed every point through the network you have made the equivalent of multiple noisy updates, where an online (passing the whole dataset through in one go) method would have only made one.

There is more advantage in mini-batching than just that though. Remember how each batch is *approximately* giving the same update as the whole data set? That noise which we now include in the optimisation is actually beneficial when we are looking at optimisations which are not strictly convex. The "jitter" imposed by the noise in each mini-batch update allows the optimisation to get out of local minima, and thus aids in finding a good solution. Additionally, there have been some studies which suggest that mini-batching also helps in finding solutions to the weight optimisation which are more generalizable. That is to say it pushes the optimisation towards flat minima, which are more likely to also be minima for the testing set. This is because by splitting up the data into mini-batches, and updating on each mini-batch, we reinforce updates which are general across mini-batches, and average away updates which are specific to one particular mini-batch, which aids us in avoiding the perennial overfitting problem.

```
def train(self, cost=tf.contrib.losses.
mean_squared_error, optimizer=tf.train.SGDOptimizer
(learning_rate=1e-3)
    , epochs=1000, batchsize=100):
```

```
"""
Train the neural network
:param cost: tf loss function
:param optimizer: tf optimizer
:return:
"""

init_all = tf.initialize_all_variables()
pred  = self._eval_train()
correct_prediction = tf.equal(tf.argmax(pred, 1),
tf.argmax(self.y, 1))
accuracy = tf.reduce_mean(tf.cast
(correct_prediction, tf.float32))
Xph = tf.placeholder(tf.float32, [None, self.X.shape
[1]])
yph = tf.placeholder(tf.float32, [None, self.y.shape
[1]])
cost = tf.reduce_mean(tf.nn.l2_loss(pred, self.y))
optimizer = optimizer.minimize(cost)
nbatches = int(self.X.shape[0] / batchsize)
with tf.Session as sess:
  for epoch in range (epochs):
    for batch in range(nbatches):
      batch_idx = np.random.randint(self.X.shape[0],
size=batchsize)
      batch_xs = self.X[batch_idx, :]
      batch_ys = self.y[batch_idx, :]
      sess.run(optimizer, feed_dict={Xph: batch_xs,
yph: batch_ys})
      train_acc = sess.run(accuracy, feed_dict={X:
batch_xs, y: batch_ys, dropout_keep_prob:1.})
```

4.5 Improving the Optimisation

4.5.1 Bias

The weight allows you to shift the slope of the activation function, but what should you use if you want instead to shift the position of the activation function? This is the job of the bias term, which contributes to the definition of

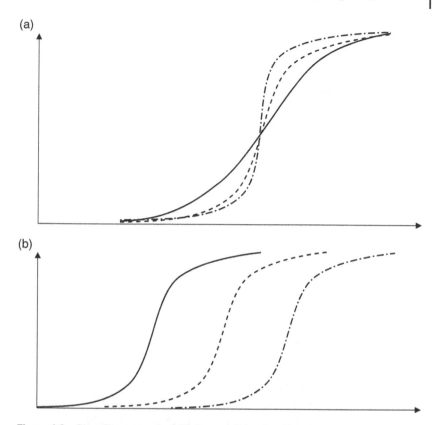

Figure 4.9 Bias allows you to shift the activation function along the X-axis. (a) We have three different sigmoid functions, but all pass through the same point at $X = 0$. When you add a bias term (b) these sigmoids separate along the X-axis.

each neural network layer. Bias allows you to shift along the X-axis, as is shown in Figure 4.9, and can be thought of as similar to the role of the intercept in linear regression ($y = mX + c$).

So why do we need bias terms? Well, without a bias term, a layer in a neural network is simply the multiplication of an input vector with a matrix. One outcome of this is that an input of all zeros will always be mapped to an output of all zeros. This might be a reasonable solution for some systems but in general it is too restrictive. Bias allows us to remove this restriction, and radically improves the generalisation and power of our networks. Let us see how this would work out in code:

```python
def _random_init_biases(self):
    """
    Randomly initialize the bias matrix
    :return:
    """
    biases = []
    for l in range(self.nlayers):
        biases.append(tf.Variable(tf.random_normal(
            [self.layer_description[l]],stddev=self.
std_init)))
        biases.append(tf.Variable(tf.random_normal(
            [self.y.shape[1]], stddev=self.
std_init)))
    self.biases = biases


def _eval_train(self, _X):
    """
    Builds the TF layers
    :return:
    """
    layers = []
    # Input layer
    layers.append(self.activation(tf.add(
        tf.matmul(_X,self.weights[0]),self.biases
[0])))
    for l in range(self.nlayers):
        layers.append(self.activation(tf.add(tf.
matmul(layers[l],
            self.weights[l+1]),self.biases[l+1])))


    # Output layer
    out = self.activation(tf.add(tf.matmul(layers
[-1],
            self.weights[-1]), self.biases[-1]))
    return out
```

4.5.2 Momentum

Training neural networks with a fixed learning rate is often slow, as the weights are modified by the same degree at each update. Momentum, introduced by Polyak et al., was designed to accelerate learning through incorporating information of previous gradients. The momentum term, μ, determines the degree to which past gradients contribute to the updates. Starting with our standard weight updates for gradient descent:

$$w_t = w_{t-1} - \eta \Delta e$$

where w_t denotes our weights at iteration t, η denotes the learning rate, and e denotes the error (or loss) associated with the network weights. To incorporate momentum, we introduce a velocity term, v, which we define as:

$$v_{t+1} = \mu v_t - \eta \Delta e(w_t)$$

Using this velocity term, we define our update as:

$$w_{t+1} = w_t + v_{t+1}$$

This has the effect of dampening oscillations in the loss function, resulting in more effective weight updates that are less prone to getting stuck in local minima. The principle here is the same as that of momentum in physics: as a ball rolls down a steep hill, it gains momentum; the more momentum it has, the less likely it is to fall into shallow ditches in the hill (local minima), and therefore the more likely it is to fall to the bottom of the hill (global minima).

4.5.3 Nesterov Momentum

Nesterov momentum is a variant of momentum that simply changes where the gradient is evaluated. Whereas in standard momentum, the gradient is computed prior to applying the velocity term, Nesterov momentum evaluates the gradient after the application of the velocity term, as in:

$$v_{t+1} = \mu v_t - \eta \Delta e(w_t + \mu v_t)$$

The weight update is then applied. This has the effect of adding a correction term to the standard momentum method, facilitating more stable gradient updates as the gradient computation incorporates the momentum term.

4.5.4 (Adaptive) Learning Rates

In addition, it may be the case that some layers learn at different rates when compared to others. Why might this be? Let us leave to one side for the moment the fact that some layers might be benefiting from a less dilute error signal and just think about the learning task. For a layer to be an effective learner, it must discover structure in the signal surface and exploit it. Unfortunately not all surfaces are well defined. Those readers who are familiar with simulation tasks such as molecular dynamics or Monte Carlo will be well aware of this problem. A layer might learn quickly when there is a single, well defined minima on the signal surface. In simulation terms you would often see this situation referred to as a funnelling landscape. This is an ideal situation for two reasons: (i) No matter where you start you end up in the minima in a small number of steps; (ii) the structure of this landscape tends to emerge very quickly. If, however, the signal surface is more complicated, the learning task is likely much harder. One reason is that you are more likely to locate local minima on the surface, and since the structure of the surface is more complicated, it is likely to evolve and change dramatically (so the local minima you find may not even be close to a minima in the next round of updates).

For those readers who are used to performing simulations, you would immediately look to treat this kind of a problem by introducing the concept of a step size. In the training of deep networks, we have a similar concept known as the learning rate, which controls the effect of the gradient on the weight updates. A large learning weight will perform harsh updates and so will train very fast, but will be unlikely to find a particularly good solution, whereas a small learning rate will train slower, and is liable to get stuck in rough parts of the landscape, but has the potential to find better overall solutions. If we again go to the world of simulations for inspiration, we might think of adaptive step size techniques, or even adaptive temperature techniques such as simulated annealing, to provide a robust solution here. Indeed, in the world of deep learning, there are a few well known adaptive step size techniques for training neural networks, and I will introduce you to them now.

4.5.5 AdaGrad

In order to overcome some of the challenges in the optimisation landscape, AdaGrad adaptively modifies the learning rate per-parameter in our model. This allows the algorithm to have higher learning rates for parameters with less-convex landscapes that are more prone to getting stuck in local minima, while also allowing parameters with smoother landscapes to reap the benefits of lower learning rates. This method of tailoring learning rates to each of the parameters allows better overall model parameters to be found, proving to be a key contribution in the field of stochastic optimisation. The formula for AdaGrad weight updates is as follows:

$$w = w - \eta \text{diag}(G)^{0.5} \circ g$$

where, G is the outer product matrix. If we look at the formula per-product updates, we have:

$$w_j = w_j - \frac{\eta}{\sqrt{G_{ij}}} g_j$$

Here, we can clearly see that, while the same learning rate (η) is used, it is adapted through multiplication with elements of the outer product matrix to produce parameter-adapted updates.

4.5.6 RMSProp

As with AdaGrad, RMSProp also adapts learning rates for each parameter. It is also one of several methods to speed up mini-batch learning. It was designed to overcome the drawbacks of the RProp optimisation technique, which tends to perform poorly for mini-batch learning as it does not average gradients over successive mini-batches – instead making substantial modifications to weights in each mini-batch, without considering the gradients of nearby mini-batches. This results in undesirable behaviour, with severe shifts in weight values causing the optimisation to "bounce" around convex structures within the objective function, without descending to the minimum. To address this, RMSProp takes a running average of the magnitudes of recent gradients for each weight. The learning rate is then divided by this value, thus moderating the gradient updates and ensuring that they are of locally similar magnitudes. This moderation of the gradient updates allows the optimisation to more effectively descend convex structures in the objective function, producing more effective weight updates:

$$G(w, t) = 0.9 \, G(w, t-1) + 0.1 \left(\frac{\delta \varepsilon(n)}{\delta w(n)} (t) \right)^2$$

4.5.7 Adam

Building on RMSProp and AdaGrad, Kingma et al. introduced the Adam method for stochastic optimisation. As with the other methods, Adam adapts the learning rate per-parameter, but in this case it builds on RMSProp's running average approach. As well as using the mean to adapt the learning rate, Adam also makes use of the variance information from the optimisation. The Adam update is defined as follows:

$$w^{t+1} = w^t - \eta \, \frac{\hat{m}_w}{\sqrt{\hat{v}_w} + \epsilon}$$

where t denotes the optimisation iteration, and $\hat{m}$ and $\hat{v}$ are the adapted mean and variance parameters defined by:

$$\hat{m}_w = \frac{m_w^{t+1}}{1 - \beta_1}$$

$$\hat{v}_w = \frac{v_w^{t+1}}{1 - \beta_2}$$

where β_1 and β_2 are used to control the rate of decay for the mean and variance parameters. The initial values for β_1 and β_2 are typically close to 1.0, for example 0.99, to facilitate a slow decay. The ϵ variable in the Adam update is simply a small noise term used to prevent division by zero in cases of zero variance. This method of using both the mean and variance information allows for more specifically tailored per-parameter updates, making Adam a particularly powerful optimiser. Given its ability to significantly improve gradient descent-based optimisation, it has become the go-to optimiser for many deep learning tasks.

4.5.8 Regularisation

When we build neural network models (or indeed any models!) we are aiming at building models which have good generalisation – that is to say they are not overfitted to the training data. One of the problems of neural network models is their inherent complexity. As you get wider and deeper networks, there will typically be many different settings of the weights that are able to reproduce the training set almost perfectly, especially if the model is trained on a small amount of data. Each of these different solutions to the learning problem will make different predictions for the validation data, and thus will all have different abilities to generalise to new data. Typically, almost all of these models will do worse on the test data than on the training data because the neurons have been tuned to work well in cohort on the training data but not on the test data. In order to reduce this problem, we must regularise the learning. There are many different ways to do this, and we will look at a few of the most commonly used.

4.5.9 Early Stopping

The premise of early stopping is pretty simple. You take a sample out of the training set and hold it out from the learning, we will call this the validation set. After every training epoch you predict the error on the validation set – if it goes up you assume that it overfitting and stop the learning.

This technique is based on a classic view of how learning progresses along the bias-variance scale (Figure 4.10):

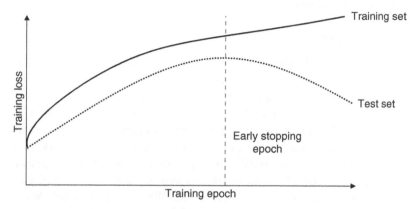

Figure 4.10 Training vs. validation error.

If your training curves look like this then it is pretty simple to see when you would stop. Unfortunately, life is rarely this simple! A more common situation looks like this (Figure 4.11):

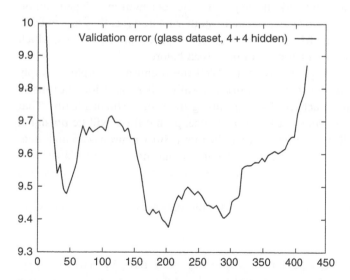

Figure 4.11 Validation error from training model on the Glass dataset. *Source:* A real validation error curve taken from Prechelt, Lutz. "Early stopping-but when?." Neural Networks: Tricks of the trade. Springer, Berlin, Heidelberg, 1998. 55–69.

In this situation, you might not want to stop if the error in the validation set rises below a certain amount. Thus, it is typical to introduce a "margin of acceptability" into the early stopping, allowing error increases of a certain amount or a certain number of consecutive increases before initialising the stopping procedure.

4.5.10 Dropout

Dropout is a method which is aimed at improving the applicability of bagging and model averaging to deep networks. The insight behind dropout is often attributed to be inspired from the role of sex in evolution. Since a child inherits half of its genes from one parent and half from another, it might be postulated that asexual reproduction is a superior means of evolving, since good combinations of genes, which are known to work well together, can be passed directly from parent to child. A very cursory glance at how most advanced organisms evolve, however, can tell you that this is not the preferred mechanism. Why might this be? Well, one interpretation is that asexual reproduction would result in large sets of genes being reliant on the presence or absence of another large set of genes in order to function well, and that this is a very poor way to evolve good "solutions."

The analogy drawn for dropout was that the same can be applied to hidden neurons in a deep network. If the performance of a neuron is dependent on the presence or absence of a signal at another neuron (i.e. they are co-evolving), then this can drive the optimisation to high-variance solutions, which are fragile to data which has not been seen before.

So how does dropout solve this? Well the premise is simple – simply make sure that whilst the network is training, it is hard for neurons to co-evolve. This is achieved by "dropping them out" stochastically. That is to say, for any neuron there is a chance, p, that there will be no signal output, regardless of the strength of the input. Since neurons can no longer rely on the output of another neuron (since it may not be there at all!), neurons evolve to be more robust to a changeable network, and thus become more independent.

One advantage to dropout is that is stunningly simple to implement Here it is in code:

```
# define our dropout array and simply multiply
# it with our weights
```

```
dropout_prob = 0.2
dropout = np.random.binomial(1, dropout_prob,
weights.shape[1])
    weights = weights * dropout
```

During dropout learning, we can expect three learning phases:

1) At the beginning of learning, when the weights are typically small and random, the total input to each unit is close to 0 for all the units and the consistency is high: the output of the units remains roughly constant across subnetworks (and equal to 0.5 with $c = 1$).
2) As learning progresses, activities tend to move towards 0 or 1 and the consistency decreases, i.e. for a given input the variance of the units across subnetworks increases.
3) As the stochastic gradient learning procedure converges, the consistency of the units converges to a stable value

4.6   Parallelisation of learning

4.6.1   Hogwild!

While it is true that Neural Networks scale far better than other methods (e.g. Gaussian Processes), they can still be slow to train in the case of large network architectures or large or high-dimensional datasets. One solution to this problem is to parallelise the mini-batch process – running each mini-batch in a separate, parallel thread. Initially, this seems like a straightforward solution – but what happens when the threads try to access memory to update the network weights? In the typical case, the memory will be locked while one thread updates the network weights, preventing other threads from updating, and producing a bottleneck that can considerably impede training.

To address this problem, a simple, yet somewhat unconventional, method was developed by researchers from the University of Wisconsin. The process removes memory locking from the equation, giving each thread access to memory and allowing each thread to update the network weights and to pull the current state of the network as and when required. This deliberate race-condition seems undesirable, but in practice conflicts are rarely encountered, as the number of weights being updated is far greater than the number of threads running the updates. In the event that conflicts do occur, their impact on overall training is negligible.

4.7 High and Low-level Tensorflow APIs

Throughout this chapter I have been relatively explicit in showing you the code at a low level, and not hiding anything in calls to pre-built functionality at least the first time I show you something. This is deliberate, as I want you to know what is going on. I do understand, however, that you do not want to be coding up your own neural network package from scratch. This is, to a degree, a waste of time, although a great educational exercise. The premise of this book is not to make you experts in coding neural network packages, it is to show you how deep learning can change the way you do research.

In the early days, TensorFlow was designed primarily with its low-level API in mind, and we make use of this occasionally to demonstrate various concepts. In these cases, we use functionality from TensorFlow version one. Luckily for everyone, TensorFlow is now designed with a key focus on its high-level API, Keras. Through Keras, TensorFlow provides a high-level, yet flexible, API that makes it easy to create a huge variety of rich neural network architectures. Keras contains classes to facilitate layer types, optimisers, and other components that are the key to design neural networks. These classes provide arguments for all of their key parameters, providing a huge degree of flexibility. The classes also provide sensible default arguments, allowing you to get started quickly, without an in-depth knowledge of the components and their underlying algorithms. Obviously the more you stick to the defaults, the more flexibility you lose, but the shorter the time between you having an idea and getting some initial results. Of course, at this point since you have a good understanding on what is going on under the hood, you are able to make some educated judgements on which parameters you may want to modify, and which you may want to leave as they are. Do you need leaky ReLUs? Hogwild parallelisation? Dropout? Maybe a more sophisticated optimiser? This is very easy to achieve through Keras. Let us go through an example of a feedforward neural network using this API as a demonstration. We will use a simple regression test case for the example, importing and configuring the data using Scikit-Learn.

```python
import numpy as np
from tensorflow import keras
from sklearn.datasets import load_boston
from sklearn.model_selection import train_test_split

# load a dataset for a simple regression problem
X, y = load_boston(return_X_y=True)

# split this into training and test data
X_train, X_test, y_train, y_test = train_test_split
(X, y, test_size=0.2)

# define our model architecture
model_arch = [
keras.layers.Dense(50, activation="relu"),
keras.layers.Dropout(0.2),
keras.layers.Dense(10, activation="relu"),
keras.layers.Dropout(0.2),
keras.layers.Dense(1)
]

# create our model using Keras
model = keras.Sequential(model_arch)

# compile our model, providing the optimizer and loss
we wish to use
model.compile(optimizer="adam", loss = keras.losses.
MeanSquaredError())

# define our early stopping callback
early_stopping = keras.callbacks.EarlyStopping
(monitor="loss", patience=3)

# train out model
model.fit(X_train, y_train, batch_size=10, epochs=50,
callbacks=[early_stopping])

# check our RMS error on the data to get an idea of our
model's performance
test_mse = model.evaluate(X_test, y_test)
rmse = np.sqrt(test_mse)
```

Here, we see that Keras provides an intuitive way of constructing our networks – dealing with them one layer at a time through their Sequential() object. We first define our architecture as a list, choosing from classes in keras.layer. In this case, we have a network comprising three fully-connected layers (Dense()), with two of them incorporating dropout, and one of them being our output layer. Once we have defined our structure within our model_arch list, we can compile our model. It is at compilation that we define two other crucial parameters for our network: the optimiser type and the loss function. Keras provides a rich variety of optimisers and loss functions to choose from. As we know Adam is particularly performant, we select this for our optimiser. As we are dealing with a simple regression problem, we use mean squared error (MSE) as our loss function.

Next, we define our early stopping criteria. To use early stopping with Keras, we need to define a callback that we pass to the fit() function – we define this here using the EarlyStopping() callback class. We set our early stopping to monitor our loss value, and to use a patience value of 3. We are now ready to train our model by calling the fit() function. We pass our training data to this, along with some key parameters – our batch size, the maximum number of epochs we want to train for, and our early stopping callback. When we run fit(), the model will train until it reaches our epochs value or training terminates due to early stopping. In this case, our model should run for around 20 iterations before early stopping terminates training. Lastly, we evaluate our model using the test data; in this case we should return a RMSE of ~10. As a follow-up exercise, why not see whether you can improve on this RMSE by changing some of our model parameters?

4.8 Architecture Implementations

Throughout the rest of the book we will explore a variety of different neural network architectures and their applications. While you will learn how to implement a variety of architectures from scratch, it is also important to point out that you can obtain high quality implementations of many architectures from various model repositories. While this won't be necessary for simpler architectures, using these sources for complex architectures can be a valuable time saver, allowing you to quickly try out sophisticated models within your work. One of the most comprehensive sources of neural network architectures is ModelZoo, which provides a rich variety of neural network implementations across all the major frameworks. You can find the Keras models (in-line with the framework we will be using throughout the book) here: https://modelzoo.co/framework/keras

4.9 Summary

Congratulations for reaching the end of our first "deep" chapter! Throughout this chapter, you have learned about the most fundamental of the deep learning models, the feedforward neural network. Now you should:

- Understand activation and loss functions
- Understand how to initialise weights in a neural network using different initialisation schemes such as random, He, and Xavier
- Understand how to update weights in a network using back propagation
- Understand the improvements to weight updates through the use of optimisers such as AdaGrad, ADAM, and RMSProp
- Understand how to regularise your learning using dropout
- Understand how to parallelise your learning using Hogwild
- Have experienced building a network using the high-level TensorFlow APIs.

4.10 Papers to Read

At the end of each chapter in this book, we will provide the interested reader with a few carefully chosen references. This is not an exhaustive list, but will highlight either a major breakthrough, or an interesting application. Here are our choices for feedforward networks:

Learning representations by back-propagating errors

– David E. Rumelhart, Geoffrey E. Hinton & Ronald J. Williams

Link: https://www.nature.com/articles/323533a0

Abstract: We describe a new learning procedure, back-propagation, for networks of neurone-like units. The procedure repeatedly adjusts the weights of the connections in the network so as to minimize a measure of the difference between the actual output vector of the net and the desired output vector. As a result of the weight adjustments, internal 'hidden' units which are not part of the input or output come to represent important features of the task domain, and the regularities in the task are captured by the interactions of these units. The ability to create useful new features distinguishes back-propagation from earlier, simpler methods such as the perceptron-convergence procedure.

Notes: This is the original paper on back propagation, and is considered to be one of the methods which made the deep learning revolution possible. There have been many updates and improvements to this algorithm over the years, but it remains a major part of how networks learn weights, and this paper is a great place to see how it all began.

Dropout: A Simple Way to Prevent Neural Networks from Overfitting

– Nitish Srivastava, Geoffrey Hinton, Alex Krizhevsky, Ilya Sutskever, Ruslan Salakhutdinov

Link: http://jmlr.org/papers/v15/srivastava14a.html

Abstract: Deep neural nets with a large number of parameters are very powerful machine learning systems. However, overfitting is a serious problem in such networks. Large networks are also slow to use, making it difficult to deal with overfitting by combining the predictions of many different large neural nets at test time. Dropout is a technique for addressing this problem. The key idea is to randomly drop units (along with their connections) from the neural network during training. This prevents units from co-adapting too much. During training, dropout samples from an exponential number of different âœthinnedâ networks. At test time, it is easy to approximate the effect of averaging the predictions of all these thinned networks by simply using a single unthinned network that has smaller weights. This significantly reduces overfitting and gives major improvements over other regularization methods. We show that dropout improves the performance of neural networks on supervised learning tasks in vision, speech recognition, document classification and computational biology, obtaining state-of-the-art results on many benchmark data sets.

Notes: Another paper including Geoff Hinton on the author list – you will see his name come up many times in the literature of deep learning as he, along with fellow pioneers Yan LeCun and Yoshua Benjio, have driven the field from the beginning. This paper shows how a seemingly simple idea such as dropout, can have an incredible impact on the training capabilities of a network. These days, it is unusual to see a network trained without including this method.

Learning from the Harvard Clean Energy Project: The Use of Neural Networks to Accelerate Materials Discovery

– Edward O. Pyzer-Knapp, Kewei Li & Alan Aspuru-Guzik

Link: https://onlinelibrary.wiley.com/doi/abs/10.1002/adfm.201501919

Abstract: Here, the employment of multilayer perceptrons, a type of artificial neural network, is proposed as part of a computational funneling procedure for high-throughput organic materials design. Through the use of state of the art algorithms and a large amount of data extracted from the Harvard Clean Energy Project, it is demonstrated that these methods allow a great reduction in the fraction of the screening library that is actually calculated. Neural networks can reproduce the results of quantum-chemical calculations with a large level of accuracy. The proposed approach allows to carry out large-scale molecular screening projects with less computational time. This, in turn, allows for the exploration of increasingly large and diverse libraries.

Notes: This paper, written by one of the authors of this textbook, demonstrates the power of using deep learning in scientific workflows, in this case materials discovery. The paper demonstrates how utilising parallel asynchronous training, alongside regularisation methods such as early stopping and multi-objective training can lead to significant savings in computational screening workflows. In this example, a trained network was able to avoid the requirement for screening 99% of a particular materials discovery workflow.

5

Recurrent Neural Networks

5.1 Introduction

Sequences are everywhere. The stock market, for example, is a sequence which charts price of a stock against time. In science, molecules can be described as a sequence of atoms, proteins can be described as a sequence of amino acids and many pieces of equipment are monitored as a sequence of measurements.

Sequences are inherently order dependent. Looking back on the multilayer perceptron (MLP) which we have worked with up to this point, it is clear that there is no concept of a sequence in how the inputs are treated – every input in a network is correlated to every other input, and once it has passed through the network, it is gone. Clearly we need to change things up a bit to let our deep learning algorithms include the concept of sequences. This is where recurrent neural networks (RNNs) come in.

5.2 Basic Recurrent Neural Networks

The first major difference between recurrent neurons and the simpler feed-forward neurons we looked at previously is that in addition to receiving the input vector, a recurrent neuron will also receive the output from the previous step.

One way to think about this is to imagine what the recurrent neuron would look like if it were viewed with respect to time. This process is known as unrolling the network through time, and when you do this it immediately becomes clear why recurrent networks preserve order. Figure 5.1 shows how the recurrent network allows information to flow through time. This

Deep Learning for Physical Scientists: Accelerating Research with Machine Learning,
First Edition. Edward O. Pyzer-Knapp and Matthew Benatan.
© 2022 John Wiley & Sons Ltd. Published 2022 by John Wiley & Sons Ltd.

Figure 5.1 A schematic of a RNN cell. *X* and *Y* are inputs and outputs, respectively, and *s* is the hidden state of the cell.

horizontal linking of neurons in time means that now the order in which the inputs are received is taken into account by the network.

We can extend this to deep networks, by thinking about recurrence using the hidden layers. In contrast to the simple flow which is shown in Figure 5.1, once we start placing recurrence into hidden layers, we become much more adept at retaining information – the hidden recurrent layers start acting like a neural memory. This allows it to apply updates in the context of the sequence seen thus far. In the case of chemistry, this could be thought of as seeing an atom in the context of its neighbours, something we know to be vitally important in determining chemical properties. In an RNN, the hidden layer at a timestep *t* is a combination of the input at *t*, and the value of the previous hidden layer.

The second difference between a recurrent network and a feedforward MLP, is that in an MLP-type network we would imagine that every timestep would have its own parameters, in a recurrent network, however, the parameters are shared across all timesteps. This has the advantage that in an RNN there are far fewer parameters, which makes training easier, and less prone to overfitting.

5.2.1 Training a Basic RNN

When we learned about training an MLP, we came across the backpropagation algorithm. This algorithm works on fixed sized inputs, and their related output – can we use it to train our network on sequences? Yes, we can – we just have to unroll the recurrent network through time during training. This strategy is called backpropagation through time (BPTT). Since weights are shared between all timesteps, we can present BPTT as follows:

1) Unroll all *N* timesteps in the network, giving us *N* sets of inputs, weights, and outputs.
2) Calculate the error and derivative for each timestep
3) Accumulate errors across the whole series
4) Roll up the network, and update the weights

From this, it is clear that this can get very expensive as *N* gets to be large, since the number of derivatives required scales with *N*. Thus, in practise we often use an approximation to BPTT known as truncated backpropagation through time (TBPTT). TBPTT takes a sequence, runs through a set number of timesteps (often referred to as k_1) and then performs BPTT over a smaller number of timesteps (often referred to as k_2). Since the number of derivatives now scales with k_2 and $k_2 < k_1$, the process is now computationally tractable. At a macro level, k_1 controls how fast and well the network trains, and is broadly equivalent to a step size, whilst k_2 controls the size of temporal structure which can be captured. To adequately train an RNN, it is important to find a balance between these two parameters.

5.2.2 Putting It All Together in TensorFlow

How might this look in code? TensorFlow has a number of pre-packaged RNN layers, so it is trivial to build and test networks. Key to our work here is the function *dynamic_rnn()* which is an operation in TensorFlow which allows us to perform dynamic unrolling in time. Unlike the (static) unrolling in time we have seen, for example in Figure 5.2, which constructs the whole graph explicitly, dynamic unrolling in time behaves much more like a while loop, iterating through the recurrent cell the appropriate number of times. This is good for two reasons:

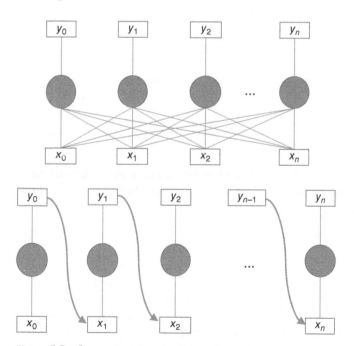

Figure 5.2 Connections in a feedforward layer in an MLP (a) destroy the sequence ordering, but a recurrent neural network layer (b, shown unrolled through time) captures it.

1) It is laborious to have to unroll everything statically – imagine writing code without while or for loops.
2) Using inbuilt functions like this allow us to pass off the optimisation work (e.g. how do I best control memory usage when I am training) to the developers of TensorFlow – who probably know much more about this than us.

Here is code for building a shallow RNN in TensorFlow, using the BasicRNN class and dynamic unrolling:

```
import tensorflow.compat.v1 as tf

# Set up some constants describing the problem
n_neurons = 100 # number of neurons
n_steps = 50 # number of timesteps
n_inputs = 150 # number of training inputs to the
network (NB this will change per problem, 150 is
arbritry here)

X = tf.placeholder(tf.float32,[None, n_steps,
n_inputs])
Y = tf.placeholder(tf.float32, [None])

shallow_rnn = tf.contrib.rnn.BasicRNNCell
(num_units=n)

output, states = tf.nn.dynamic_rnn(shallow_rnn, X,
dtype=tf.float32)
```

Of course, we want to see if this can work on a deep network, with many recurrent layers. Fortunately, this is easy in TensorFlow, using the *Multi-RNNCell* class.

```
Import tensorflow.compat.v1 as tf

# Set up some constants describing the problem
n_neurons = [100, 50, 25] # number of neurons
n_steps = 50 # number of timesteps
n_inputs = 150 # number of training inputs to the
network (NB this will change per problem, 150 is
arbritry here)
```

```
X = tf.placeholder(tf.float32,[None, n_steps,
n_inputs])
Y = tf.placeholder(tf.float32, [None])

rnn_layers = []
for n in n_neurons:
    cell = tf.contrib.rnn.BasicRNNCell(num_units=n)
    rnn_layers.append(cell)

deep_rnn = tf.contrib.rnn.MultiRnnCell(rnn_layers)

output, states = tf.nn.dynamic_rnn(deep_rnn, X,
dtype=tf.float32)
```

5.2.3   The Problem with Vanilla RNNs

One of the major problems with vanilla RNNs occurs during the propagation of errors back through the network. Since the recurrent behaviour of the network means that the weight matrix connecting the input to the hidden layer is multiplied a large number of times (the number of timesteps used, in fact), if the leading eigenvalue of the weight matrix is smaller than one, then the gradients will quickly vanish to zero. Conversely, if the leading eigenvalue of the weight matrix is greater than one, the magnitude of the elements in the weight matrix will quickly explode. Both of these behaviours have a significant and detrimental effect upon learning and make it very hard to learn long-term dependencies within the data set.

Secondly, the neural memory concept starts to break down in "vanilla" RNNs when training over many timesteps. Let us consider a tiny network with four recurrent neurons in a hidden layer. Figure 5.2 shows how the information from the initial input fades away as time progresses.

In this example, the network will begin to forget after the fifth member of the sequence is ingested (Figure 5.3). At this point, the network has to choose what to remember and what to forget. For both of these problems we need a way to capture longer term temporal relationships in the network. Variants on the vanilla RNN, the long short-term memory (LSTM) and gated recurrent unit (GRU) networks include a more sophisticated update mechanism to provide us with just this, and are widely used in place of vanilla RNNs, which are rarely used in the "real world" today.

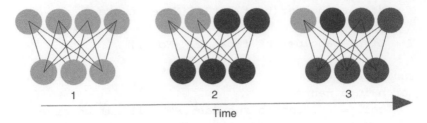

Time

Figure 5.3 An example of how sequential information is stored in a recurrent neural network. It can be seen that the hidden layer contains information from each of the previous timesteps. Since there are only four neurons in the hidden layer, when the fifth timestep is reached, the neural memory will be full and it will have to decide what to forget.

5.3 Long Short-Term Memory (LSTM) Networks

LSTM cells were developed in the late 1990s to solve the problems of the vanilla RNNs. From a practical point of view, LSTMs can be slotted in whenever you are thinking of using a vanilla RNN, and will improve performance and training speed. It's not a great idea, however, to just blindly slot in something without at least having a reasonable idea why (why else would you be reading this book?), so let's take a bit of a deeper dive. When you read about LSTMs, you will often see a diagram representing the flow of information, which looks something like this (Figure 5.4):

This looks a little intimidating, especially compared to the relatively simple neurons and cells we have looked at thus far, so we will go ahead and think about this in more detail.

5.3.1 Forget Gate

One of the key differences between an LSTM cell and a vanilla RNN cell is the presence of a so-called forget gate (Figure 5.5). This gate controls how much of the memory at $t - 1$ is kept, and how much is forgotten – this is the mechanism which allows the LSTM to understand longer term trends.

The forget gate is fed by all three inputs, although they do not all take the same route to get there. The memory input reaches direct from input to forget gate – this can be thought of as a flow of purely "old" memories. Of course, we need to introduce new information to the memory flow. How these two flows are combined is controlled by a structure known as a gate. A gate is a combination of a one layer sigmoidal neural network and an elementwise multiplication operation. The first gate we will meet is the forget gate which controls the mixing of the two information flows.

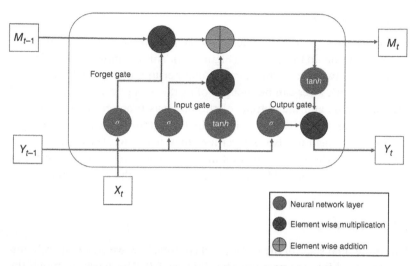

Figure 5.4 A schematic of information flow through an LSTM cell. As throughout this book, X refers to cell inputs, Y, outputs, and M in this case refers to the memory state of the cell.

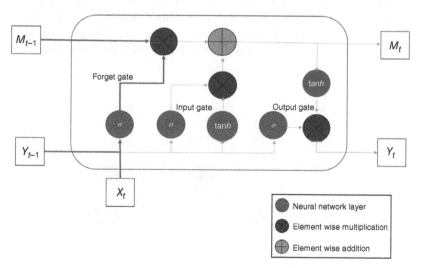

Figure 5.5 An LSTM cell with the flow through the forget gate highlighted.

The amount of the old memory which is kept is controlled by the output of the sigmoidal layer, which has a value for each element in the memory. If the value is 0, the cell completely forgets the old input, whereas if it is a 1 it is completely kept.

5.3.2 Input Gate

The role of the input gate is to decide what parts of the new information we are going to store in the memory (Figure 5.6). This is controlled by a sigmoidal neural network layer in a similar manner to the forget gate. The tan$h$ neural network layer can be thought of as the guts of a vanilla RNN, as it uses the inputs and the outputs of the previous timestep to create a new memory state. This is then multiplied by the output of the input gate layer (the sigmoidal neural network layer next to it) to create a vector of new candidate values. These are then combined with the output of the forget gate using the elementwise addition operation to create the new memory.

5.3.3 Output Gate

This is the final stage of the LSTM, and controls how we create the cell outputs to be fed into the next timestep (Figure 5.7). This works in much the same way as previous gates, in so far as how the inputs and outputs of the previous timestep are combined is controlled by a sigmoidal neural network layer. For this final step, we want to reintroduce the memory of the cell, which we have pushed through a tan$h$ activation to make the values sit within the range $-1,1$. These two streams are combined at the output gate through an elementwise multiplication to create the output of the cell.

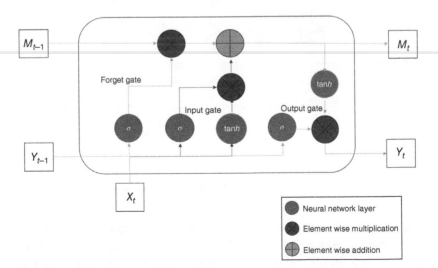

Figure 5.6 An LSTM cell with the flow through the input gate highlighted.

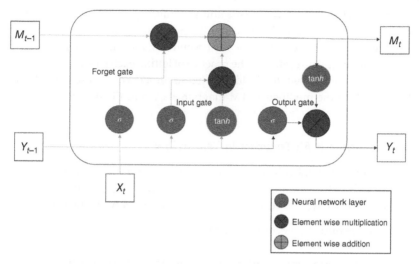

Figure 5.7 An LSTM cell with the flow through the output gate highlighted.

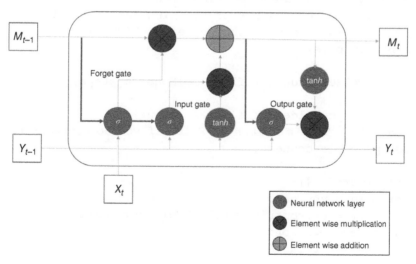

Figure 5.8 An LSTM cell with peephole connections highlighted.

5.3.4 Peephole Connections

As with most things in deep learning, there is no single variant of the LSTM (Figure 5.8). What you have just worked through covers how the majority of LSTMs are implemented, but of course researchers have changed up the connections inside the LSTM to see how this affects performance. One of the most important of these changes, and the one which you are most likely to come across in the wild, is the so-called peephole connection. This is actually pretty simple, and just refers to the practice of letting each gate have a "peep" at the cell's memory whilst deciding what to forget. You can see how these flows work in the modified LSTM diagram shown in Figure 5.8.

5.3.5 Putting It All Together in TensorFlow

Now you understand LSTMs, let us look at how you might build a simple LSTM-RNN network using the TensorFlow application programming interface (API). Luckily, it is very similar to the basic RNN we looked at earlier.

```
import tensorflow.compat.v1 as tf

# Set up some constants describing the problem
n_neurons = [100, 50, 25] # number of neurons
n_steps = 50 # number of timesteps
n_inputs = 150 # number of training inputs to the
network (NB this will change per problem, 150 is
arbritry here)

X = tf.placeholder(tf.float32,[None, n_steps,
n_inputs])
Y = tf.placeholder(tf.float32, [None])
peephole = True # Switch to False if you do not want to
use peephole connections

rnn_layers = []
for n in n_neurons:
    cell = tf.contrib.rnn.LSTMCell(num_units=n,
use_peephole=peephole)
    rnn_layers.append(cell)

deep_rnn = tf.contrib.rnn.MultiRnnCell(rnn_layers)

output, states = tf.nn.dynamic_rnn(deep_rnn, X,
dtype=tf.float32)
```

5.4   Gated Recurrent Units

Recently, GRUs have become a popular alternative to RNNs and LSTMs. At a high level, GRUs are a simpler version of an LSTM which display similar behaviours and performance (which is perhaps why they are popular!).

Figure 5.9 shows how information flows through a GRU cell. As you can see, it is similar to an LSTM cell, but there are also some marked differences. Importantly, there is one fewer gate – a GRU does not have an output gate, instead returning the entire state vector at every timestep. You might also notice that there is no separate memory flow through this cell with the M and Y inputs from the LSTM being replaced with a single H input here. These simplifications mean that there are fewer parameters to tune, and consequently, GRU cells are often faster to learn and can potentially generalise better since there is lesser chance of overfitting. Their relative simplicity when compared to LSTMs, however, may lead to them being unable to express certain complex trends. As you will see, though, when you use the higher level functionality of TensorFlow, it is easy to build both and see which suits your problem better.

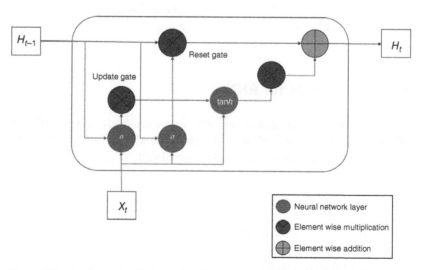

Figure 5.9   A schematic of information flow through a GRU cell. Here, X refers to cell inputs and H the combined memory and output state of the GRU.

5.4.1 Putting It All Together in TensorFlow

```
import tensorflow.compat.v1 as tf
# Set up some constants describing the problem
n_neurons = [100, 50, 25] # number of neurons
n_steps = 50 # number of timesteps
n_inputs = 150 # number of training inputs to the
network (NB this will change per problem, 150 is
arbritry here)

X = tf.placeholder(tf.float32,[None, n_steps,
n_inputs])
Y = tf.placeholder(tf.float32, [None])

rnn_layers = []
for n in n_neurons:
    cell = tf.contrib.rnn.GRUCell(num_units=n)
    rnn_layers.append(cell)

deep_rnn = tf.contrib.rnn.MultiRnnCell(rnn_layers)

output, states = tf.nn.dynamic_rnn(deep_rnn, X,
dtype=tf.float32)
```

5.5 Using Keras for RNNs

Previous examples have shown how we can construct RNNs with different cell types using the TensorFlow 1 API. Here, we will see how easily RNNs can be constructed using the Keras Sequential API:

```
From tensorflow import keras

# define our network architecture
model_arch = [
keras.layers.LSTM(64),
keras.layers.Dropout(0.1),
keras.layers.Dense(o_len)
]
```

```
model = keras.Sequential(model_arch)

# compile and train our network
model.compile(loss="mean_squared_error",
              optimizer=tf.optimizers.Adam(),
              metrics=["mean_squared_error"])

model.fit(X_train, y_train,
          batch_size=32,
          epochs=100,
          validation_split=0.2)
```

Just as with the LSTM example here, we can easily use GRU cells by simply using keras.layers.GRU – making it incredibly straightforward to create a variety of RNN networks.

5.6 Real World Implementations

Several useful pre-implemented RNN architectures can be found in Model-Zoo, with examples of how these are used in real-world applications:

1) DeepSpeech2: an end-to-end automatic speech recognition model using bidirectional LSTMs. Link: https://modelzoo.co/model/deepspeech
2) Image Compression with Neural Networks: this uses RNNs with GRU cells to compress images. While this is useful for image compression, the architecture could also be adapted for other data compression tasks. Link: https://modelzoo.co/model/compression

How might you use these architectures within your own work?

5.7 Summary

In this chapter we have seen how RNNs employ recurrence to effectively model sequential information. You should now have a good understanding of:

• Traditional RNNs, and the use of recurrence in creating predictive models
• Shortcomings of RNNs: exploding and vanishing gradients

- LSTM and GRU cells, their components, and how they overcome short-comings of traditional RNNs
- How to implement various types of RNNs using both the high- and low-level TensorFlow APIs.

5.8 Papers to Read

Here we present a number of useful resources for learning more about the development of RNNs and their use within a variety of applications.

A Novel Connectionist System for Unconstrained Handwriting Recognition

– Alex Graves *et al.*, IEEE Transactions on Pattern Recognition and Machine Intelligence, 2009

Link: https://dl.acm.org/doi/10.1109/TPAMI.2008.137

Abstract: Recognizing lines of unconstrained handwritten text is a challenging task. The difficulty of segmenting cursive or overlapping characters, combined with the need to exploit surrounding context, has led to low recognition rates for even the best current recognizers. Most recent progress in the field has been made either through improved preprocessing or through advances in language modeling. Relatively little work has been done on the basic recognition algorithms. Indeed, most systems rely on the same hidden Markov models that have been used for decades in speech and handwriting recognition, despite their well-known shortcomings. This paper proposes an alternative approach based on a novel type of recurrent neural network, specifically designed for sequence labeling tasks where the data is hard to segment and contains long-range bidirectional interdependencies. In experiments on two large unconstrained handwriting databases, our approach achieves word recognition accuracies of 79.7 percent on online data and 74.1 percent on offline data, significantly outperforming a state-of-the-art HMM-based system. In addition, we demonstrate the network's robustness to lexicon size, measure the individual influence of its hidden layers, and analyze its use of context. Last, we provide an in-depth discussion of the differences between the network and HMMs, suggesting reasons for the network's superior performance.

Notes: This is one of the landmark RNN papers that demonstrated their ability to greatly improve on traditional machine learning methods for sequence/context-based tasks. In this work, Alex Graves et al. demonstrate how RNNs can be used to greatly improve handwriting recognition systems. Prior to this work, handwriting recognition methods were poor, often based on hidden Markov models (HMMs) – relatively simple statistical models that are n't able to effectively model contextual features in the data. Graves' work demonstrates that LSTM-RNNs are able to overcome the shortcomings of traditional methods, producing significant improvements over the then-state-of-the-art.

Speech recognition with deep recurrent neural networks

– Alex Graves, *et al.,* IEEE International Conference on Acoustics, Speech, and Signal Processing

Link: https://arxiv.org/abs/1303.5778

Abstract: Recurrent neural networks (RNNs) are a powerful model for sequential data. End-to-end training methods such as Connectionist Temporal Classification make it possible to train RNNs for sequence labelling problems where the input-output alignment is unknown. The combination of these methods with the Long Short-term Memory RNN architecture has proved particularly fruitful, delivering state-of-the-art results in cursive handwriting recognition. However RNN performance in speech recognition has so far been disappointing, with better results returned by deep feedforward networks. This paper investigates \emph{deep recurrent neural networks}, which combine the multiple levels of representation that have proved so effective in deep networks with the flexible use of long range context that empowers RNNs. When trained end-to-end with suitable regularisation, we find that deep Long Short-term Memory RNNs achieve a test set error of 17.7% on the TIMIT phoneme recognition benchmark, which to our knowledge is the best recorded score.

Notes: Another landmark RNN paper by Graves, this time working with Geoffrey Hinton, demonstrating how *Deep* LSTM-RNNs can be used to achieve state-of-the-art performance in speech recognition tasks. While RNNs were investigated for speech recognition tasks prior to this work, they achieved poor results – often being outperformed by simpler deep feedforward networks. Furthermore, much of the earlier work on speech recognition relied on applying transforms to the acoustic data in order to create features of reduced dimensionality that could be processed by traditional methods (often HMMs). In this work, Graves et al. demonstrate how deep bidirectional LSTM-RNNs (BLSTMs) can be used to learn directly from the acoustic information, using multiple BLSTM layers in order to create progressively high-level representations of the data. This method produced significant improvements over state-of-the-art techniques, and still forms the core of many speech recognition systems today.

How to Construct Deep Recurrent Neural Networks

– Razvan Pascanu *et al.*, Proceedings of the International Conference on Learning Representations, 2014

Link: https://arxiv.org/abs/1312.6026

Abstract: In this paper, we explore different ways to extend a recurrent neural network (RNN) to a \textit{deep} RNN. We start by arguing that the concept of depth in an RNN is not as clear as it is in feedforward neural networks. By carefully analyzing and understanding the architecture of an RNN, however, we find three points of an RNN which may be made deeper; (1) input-to-hidden function, (2) hidden-to-hidden transition and (3) hidden-to-output function. Based on this observation, we propose two novel architectures of a deep RNN which are orthogonal to an earlier attempt of stacking multiple recurrent layers to build a deep RNN (Schmidhuber, 1992; El Hihi and Bengio, 1996). We provide an alternative interpretation of these deep RNNs using a novel framework based on neural operators. The proposed deep RNNs are empirically evaluated on the tasks of polyphonic music prediction and language modeling. The experimental result supports our claim that the proposed deep RNNs benefit from the depth and outperform the conventional, shallow RNNs.

Notes: This paper provides some useful insights into the construction of deep RNNs.

6

Convolutional Neural Networks

6.1 Introduction

Many of the popular examples of deep learning's success incorporate convolutional neural networks (CNNs). Most obviously, they have revolutionised the field of Computer Vision, and underlie many popular image processing technologies in use today. These technologies include face detection, handwriting recognition, and lesion classification within radiology. Image processing tasks such as these require a multi-tiered approach, incorporating both local and global information to identify complex patterns. Traditional computer vision approaches accomplished this through segmenting images and applying a variety of transforms to the data at different stages. The power of CNNs lies in their ability to effectively learn how to transform data at multiple resolutions – resulting in sophisticated, but compact, representations of the data. This makes it possible to tackle complex image processing tasks with a single algorithm, and removes the need for complicated feature engineering. While they initially gained popularity due to the impressive results achieved in Computer Vision tasks, CNNs have proven to be incredibly powerful across a range of domains, including molecular representation, genomics sequence recognition, and language modelling.

In their 2015 paper, David Duvenaud et al. demonstrated that CNNs could be applied directly to graph data. They applied this approach to molecular feature extraction to produce "neural fingerprints." These neural fingerprints exhibited better performance for a range of tasks, including predicting solubility, drug efficacy, and photovoltaic efficiency.

CNNs have also proven to be a powerful tool for processing genetics data. Lyu and Haque's work from 2018 demonstrated that CNNs can obtain competitive performance for tumour type classification when applied to gene expression data. In this case, the gene expression data is first converted to

Deep Learning for Physical Scientists: Accelerating Research with Machine Learning,
First Edition. Edward O. Pyzer-Knapp and Matthew Benatan.
© 2022 John Wiley & Sons Ltd. Published 2022 by John Wiley & Sons Ltd.

an image before being processed by the CNN. A similar method is used by Chieng et al. in their 2017 work on modelling DNA sequences. In this work, the authors propose converting one-dimensional DNA sequences to a two-dimensional matrix by first representing nucleotides with numerical values, following which the one dimensional nucleotide sequences are simply reshaped into $m \times n$ matrices. Using this data with a three-layer two-dimensional CNN produced strong results in DNA motif discovery tasks, and was more computationally efficient than competing methods.

In this chapter we will introduce the fundamental principles of CNNs, and get to grips with using TensorFlow for both traditional CNNs and graph convolutional neural networks, or GCNNs.

6.2 Fundamental Principles of Convolutional Neural Networks

6.2.1 Convolution

CNNs get their name from their use of convolution layers, which use convolution to learn a feature map from the input data. Convolution is a mathematical function that is used in place of matrix multiplication in the convolutional layers of a CNN. The convolution operation operates on two functions, f and g, and is denoted with an asterisk:

$$s(t) = (f*g)(t) \tag{6.1}$$

Convolution describes the integral of the product of the two functions after one is reversed and shifted, such that:

$$(f*g)(t) = \int f(x)g(t-x)\mathrm{d}x \tag{6.2}$$

In the case of CNNs, we refer to f as the input, g as the kernel, and our output $s(t)$ as the feature map. As we are computationally limited to discrete data, we actually use a discrete convolution – adapting Eq. (6.2) to:

$$(f*g)(t) = \sum_{x=0}^{\infty} f(x)g(t-x) \tag{6.3}$$

Commonly, CNNs are applied to two-dimensional data, such as images. In this case, we use a two dimensional input I and kernel K:

$$(I*K)(i,j) = \sum_{m}\sum_{n} I(m,n)K(i-m,j-n) \tag{6.4}$$

CNNs are also often applied to three-dimensional data, but theoretically can be applied to any number of dimensions – however, increasing dimensionality also significantly increases computational cost, and makes CNNs progressively more expensive and difficult to train.

In the case of CNNs, the input I is multidimensional array of data, and the kernel K is a multidimensional array of parameters that are learned through backpropagation. These kernel parameters are analogous to the neurons in a typical MLP layer.

6.2.2   Pooling

Another important component of CNNs is the pooling layer, which typically follows a convolution layer, resulting in a stacked architecture as shown in Figure 6.1. The pooling layer acts a kind of down-sampling or compression of the feature map produced by the preceding convolution layer. This is achieved through applying a sliding window, often referred to as the locally receptive field (LRF), to the feature map and computing some summary statistics for the contents of the window. Typically, the maximum or the mean of the window are the summary statistics used, resulting in two key types of pooling: max pooling and average pooling, which we will discuss in more detail later. In the two-dimensional case, this involves applying an $m \times n$ sliding window to the feature map. The size of the window is an important hyper-parameter that determines the degree of down-sampling.

6.2.2.1   Why Use Pooling?

In most tasks for which CNNs are used, the convolution layers produce very large feature maps – and the task of training a model with such high-dimensional inputs is both computationally demanding and prone to over-fitting. The pooling layers therefore provide a mechanism by which the dimensionality can be reduced, resulting in more efficient training and better model generalisation. Furthermore, in many applications for which

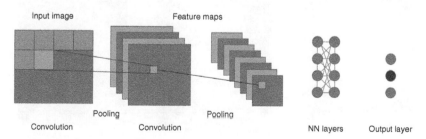

Figure 6.1   Illustration of convolutional neural network architecture.

CNNs are used, spatially local structures often recur, but vary in orientation. Without the pooling layer, these varying translations of similar structures appear as a complex combination of distinct structures – making the task of learning from these features significantly more difficult. Learning these complex combinations of seemingly distinct structures is also a key cause of the overfitting mentioned previously. The summary statistic-driven down-sampling provided by the pooling layer results in a property known as *translation invariance*. This allows similar local structures to be represented by the same (or similar) values in the pooling layer, irrespective of their orientation. In this way, pooling greatly improves the efficiency of the representation, making it easier to train the model, and resulting in better generalisation (Figures 6.2 and 6.3).

6.2.2.2 Types of Pooling

6.2.2.2.1 Average Pooling For average pooling, the average – typically the mean – of the LRF is computed, as demonstrated on the left of Figure 6.4.

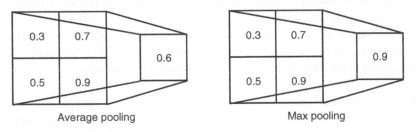

Average pooling   Max pooling

Figure 6.2 Illustration of average and max pooling algorithms.

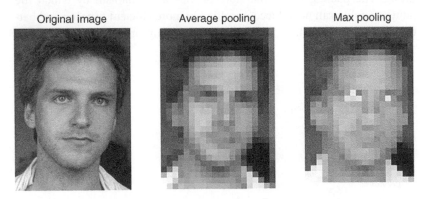

Original image   Average pooling   Max pooling

Figure 6.3 Illustration of average and max pooling on face image.

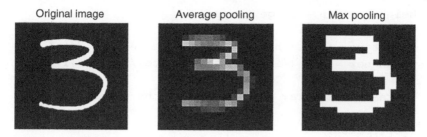

Figure 6.4 Illustration of average and max pooling on handwritten character image.

Average pooling has a smoothing effect on the data and can be useful when general information is informative for the task, and extreme local information could be problematic. For example, we may care more about the general shape than distinct local features, and so down-sampling in such a way that local features are not emphasized (but the general shape is preserved) is beneficial. This is particularly useful with noisy data, where average pooling will help to reduce the impact of the noise. Using average pooling will help to add robustness to noise, and will focus the model on broader, more general features – something that is particularly useful for a range of computer vision tasks, e.g. face detection. As we can see in the example in Figure 6.4, while the image has been significantly down-sampled, the use of average pooling has preserved much of the general information. In the case of max pooling, general information has been lost while extrema in the data – such as reflection of light off of the eyes – are emphasized. Let us take a look at what average pooling looks like in code:

```
import numpy as np
from matplotlib.image import imread
import matplotlib.pyplot as plt

image = imread('example_image.png')

# Set the pooling parameters - our kernel size and
stride.
# We assume that we want our kernel height and width to
be the same,
# producing a kernel of shape (kernel_size,
kernel_size)
kernel_size = 2
stride = 2
```

```
height, width = image.shape

height_new = int((height - kernel_size)/stride)+1
width_new = int((width - kernel_size)/stride)+1

# Create the array that will hold the data from the
pooling
# operation.
pool_output = np.zeros((height, width))

# Loop over our data, iterating by our stride length
in each
# dimension
y_origin = 0
y_out = 0
while y_origin + kernel_size
    x_origin = 0
    x_out = 0
    while x_origin + kernel_size
        y_end = y_origin + kernel_size
        x_end = x_origin + kernel_size
        pool_output[y_out, x_out] = np.mean(image
[y_origin:y_end, x_origin:x_end])
        x_origin += stride
        x_out += 1
    y_origin += stride
    y_out += 1
```

```
# Plot our pooled data
plt.imshow(pool_output, cmap='gray')
```

6.2.2.2.2 *Max Pooling* For max pooling, the maximum of the LRF is computed, as demonstrated on the right of Figure 6.4. Max pooling is useful for emphasizing information about extrema in local regions and can therefore be useful in cases where extreme values are informative. This will help the model to pick up on these features as they are emphasized by the pooling layer. Max pooling is therefore useful in tasks where distinguishing key shapes is important, such as in optical character recognition. As we can see in Figure 6.4, max pooling allows us to significantly down-sample the image while preserving – and helping to emphasize – useful information.

In contrast, for this example average pooling distorts key information in the image, resulting in a poorer representation that would make training the model more difficult. The code for this is almost identical to the previous example of average pooling, however in this case we replace the call to "np.mean()" with a call to "np.max()."

6.2.2.2.3 Global Pooling Global pooling down samples the entire feature map, or input, to a single value. This is the equivalent of using a typical pooling layer and setting the pool size to the size of the input. This is a particularly severe form of down sampling and is useful for cases in which significantly compressing information may be useful, for example creating a specific feature map for each classification output of an image classifier. For this reason, global pooling layers are sometimes used as an alternative to fully connected layers.

6.2.3 Stride and Padding

Stride and padding are two important parameters that define how our CNN layers are applied to the data (Figure 6.5). Stride is simply the amount that we shift the kernel in a given dimension each time we move the kernel. The larger the stride, the greater the difference in size between our input data and our output data. For example, if we apply a 2×2 kernel to an 8×8 matrix with a stride of 1, we will end up with an output of size 7×7. If we increase this to a stride of 2, we will end up with an output of size 4×4. As well as down-sampling, stride can be helpful for reducing computational cost, as it can dramatically reduce the number of convolution/pooling operations we need to execute per layer – but we should be wary of the type of data we are processing and its initial resolution when considering the stride value: too large a stride can result in the kernel skipping over important information in our data (Figure 6.6).

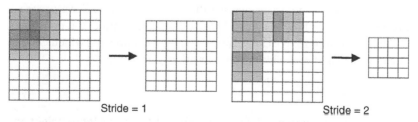

Stride = 1   Stride = 2

Figure 6.5   Illustration of the effect of stride on change in data volume.

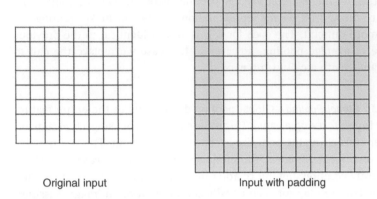

Original input   Input with padding

Figure 6.6   Illustration of stride.

While down-sampling can be computationally advantageous, down-sampling too aggressively in the early layers of the network will result in rapid loss of information – particularly information pertaining to low-level features. Padding helps to preserve low-level information across layers by reducing the down-sampling effect produced by convolution and pooling. It also improves extraction of lower-level features by allowing the data to be processed more comprehensively by the kernel. Without padding, the data at the edges of the matrix are processed fewer times than the data on the inside. Adding padding facilitates the extraction of more information from these regions by ensuring that data at the edges are processed equivalently to data elsewhere in the input. If we start with an input of size 16×16 and have a kernel of size 3×3 and use a stride of 1, we can compute the padding necessary to preserve our input size using:

$$P = \frac{K - S}{2} \tag{6.5}$$

where K is the size of our kernel, S is our stride, and P is the resulting padding parameter. In this case, we see that $P = 1$, meaning if we process a matrix of size 17×17 with a kernel of size 3×3 and a stride of 1, our output will be of size 16×16. We can verify this using the following, which gives us our output height and width given our other parameters:

$$O_{hw} = \frac{I_{hw} - K + 2P}{S} + 1 \tag{6.6}$$

where O_{hw} is our output height and width and I_{hw} is our input height and width. *Note – obviously this equation is only useful in cases for which we are processing square matrices.*

6.2.4 Sparse Connectivity

CNN's are particularly efficient in comparison to other neural network architectures thanks to two key properties: sparse connectivity and parameter sharing. Sparse connectivity is achieved through the CNN's use of a kernel, rather than a typical weight matrix used by a traditional neural network. The kernel, while still made up of weights, is typically far smaller than the input, as it is applied as a sliding window over the input, rather than having a weight associated with each input feature. This means every input unit $x_n \in X$ only affects k output units, where k is the width of the kernel. Likewise, every output $o_n \in O$ is only affected by k input units. This sparse connectivity to the output units gives rise to the concept of a unit's *receptive field*, and this grows the deeper into the network the units are. This means that all necessary information from the input can be passed through the network via indirect connections, despite the fact that all direct connections are sparse. This is demonstrated in Figure 6.7 – here we have an input of size 9×9 and two convolutional layers that each use kernels of size 3×3 (for simplicity, we assume that we use appropriate padding to preserve our original data volume). As we see here, while a unit in the first layer only has a receptive field of size 3×3, a unit in the second layer has a receptive of size 9×9 – in this case incorporating information from the entire input. Computationally, sparse connectivity results in significant gains: a typical network with input of size m and output of size n will have time complexity $O(m \times n)$, whereas, thus using a kernel of length k will have time complexity $O(k \times n)$, which, given that k is typically several orders of magnitude smaller than m, making the CNN significantly more efficient than a typical neural network.

6.2.5 Parameter Sharing

The second property that contributes the efficiency of CNNs is parameter sharing, and is again due to the kernel. Whereas a traditional neural

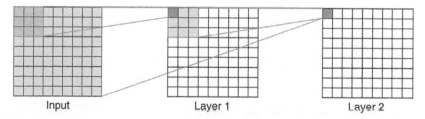

Input   Layer 1   Layer 2

Figure 6.7 Illustration of the impact of sparse connectivity on CNN unit's receptive field.

network uses each weight in the weight matrix only once (when processing the corresponding element of the input), a CNN applies the same kernel to each element of the input, therefore reusing the same set of weights. This has a dramatic effect on the memory requirements of the network, reducing this from $m \times n$ parameters to k parameters. As k is typically orders of magnitude smaller than m, this makes CNNs significantly more memory efficient than their traditional counterparts.

6.2.6   Convolutional Neural Networks with TensorFlow

While CNNs are useful for a variety of applications and data types, one of the most intuitive ways to start understanding them is through applying them to image data. Given this, the example here demonstrates how to apply a CNN to MNIST – a popular dataset for machine learning (ML) benchmarking which comprises images of handwritten digits. To put together what we have learnt in this chapter, we will use TensorFlow's high level API Keras, as this has well optimised implementations of each of the CNN components.

```
import tensorflow as tf
from tensorflow.keras import datasets, layers, models

(X_train, y_train), (X_test, y_test) = datasets.
mnist.load_data()

X_train = X_train.reshape(X_train.shape[0], 28, 28, 1)
X_test = X_test.reshape(X_test.shape[0], 28, 28, 1)

model = models.Sequential()
model.add(layers.Conv2D(28, kernel_size=(3, 3),
input_shape=(28, 28, 1)))
model.add(layers.MaxPooling2D(pool_size=(2, 2),
strides=2, padding="same"))
model.add(layers.Conv2D(28, kernel_size=(3, 3),
activation='relu'))
model.add(layers.AveragePooling2D(pool_size=(2, 2)))
model.add(layers.Flatten())
model.add(layers.Dense(64, activation='relu'))
model.add(layers.Dense(10, activation='softmax'))

model.compile(optimizer='adam',
              loss='sparse_categorical_crossentropy',
              metrics=['accuracy'])
```

```
model.fit(X_train, y_train, epochs=10)

loss, acc = model.evaluate(X_test, y_test)
print(acc)
```

As we can see from the accuracy of ~98%, our model is quite effective for classifying numerical characters from image data. Notice that in Tensor-Flow, switching between max or average pooling involves the use of the MaxPooling2d() layer or the AveragePooling2D() layer, respectively. Another thing to note is that we do not need to explicitly define our padding – if we want to preserve the shape of our data after the convolution or pooling operations, we simply pass the argument "padding='same'," and TensorFlow will work out how much padding is required and add this to the input. Lastly, TensorFlow has conveniently optimised the convolution and pooling operations for one, two, and three dimensions – switching between these simply involves changing the suffix to "1D," "2D," or "3D," respectively.

6.3 Graph Convolutional Networks

Many types of data are represented as graphs – from social networks through to molecular structure data. This has given rise to another powerful convolution-based network – the graph convolutional network, or GCN. Like CNNs, GCNs also use a kernel, or filter, to process spatially local information in the graph. In the case of GCNs, these filters are centred on a signal node of the graph, processing the node and its neighbours. However, unlike CNNs, which operate on an N-dimensional grid (and thus each filter operation always operates on the same number of elements), GCNs need to be able to process varying numbers of elements for each filter operation. This is because individual nodes in a graph – unlike pixels in two dimensional images – will not always have the same number of neighbours.

If we represent a graph as $G = (V, E)$, comprising vertices (or nodes) V and edges E, whereby an individual node is represented by $v_i \in V$, we can represent the neighbours of a given node as $N(v_i)$. Within a convolutional layer of the GCN, we want to aggregate the information of each node v_i by incorporating its own features, which we represent by x_v, and its neighbour's features, represented by x_u where $u \in N(v_i)$. As in a standard CNN, the aggregated GCN features for each node are passed through an activation function. Also like a standard CNN, GCN's make use of pooling layers.

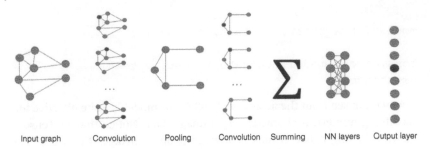

Figure 6.8 Illustration of graph convolutional network.

Where CNN pooling layers smooth or compress the data, pooling in GCN's performs a similar operation, resulting in a compressed representation of the graph. This is illustrated in Figure 6.8.

As with a standard CNN, it is common to stack multiple convolution and pooling layers. For predictive tasks, such as classification or regression, it is common to incorporate a summing layer, followed by multiple neural network layers (often fully connected multi-layer perceptron layers), before a final output layer. In the case of Figure 6.8, the GCN is being used for classification – in this case the output layer would typically be preceded by a softmax activation function.

6.3.1 Graph Convolutional Networks in Practice

For our GCN example we'll build a binary classifier to classify drug toxicity using data from the Tox21 dataset. This can be obtained from: https://tripod.nih.gov/tox21/challenge/data.jsp

The molecular data comes in the SMILES format, and we can use the Python library RDKit (https://www.rdkit.org/) to load this data.

Raw SMILES data consists of strings which describe the atoms and bonds of each molecule. While this is a convenient representation, it is not a representation that can be effectively processed by convolutions. RDK it allows us to convert this to a representation that will have the spatial properties necessary for convolutions to be effective. This representation is the adjacency matrix – a matrix representing the relationships between different nodes in the graph. For example, say we have the graph (Figure 6.9):

From this, we can construct the adjacency matrix. Each row corresponds to a node, and each column corresponds to that node's connections to other nodes – if the node at row i is connected to the node at column j, a 1 is entered into the matrix. If there is no connection, a 0 is entered, for example:

Figure 6.9 Example graph.

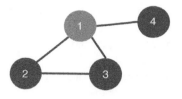

Figure 6.10 Example adjacency matrix.

Node				
1	0	1	1	1
2	1	0	1	0
3	1	1	0	0
4	1	0	0	0

As shown in Figure 6.10, we can clearly see that node 1 is connected to nodes 2, 3, and 4 – which we can confirm by looking at the graph illustration in Figure 6.10. The following code provides an example of how we can load molecular graphs, obtain the adjacency matrices, and train a GCN on this data.

```
from tensorflow.keras import datasets, layers, models
from sklearn.model_selection import train_test_split
import numpy as np

from rdkit.Chem import MolFromSmiles
from rdkit.Chem import AllChem as ac
import networkx as nx

datafile = "nr-er.smiles"
f = open(datafile, "r")
data = f.readlines()
f.close()

feats = []
X = []
y = []

# Get adjacency matrix from each SMILES molecule and
the label
# indicating whether it's toxic (1) or non-toxic (0)
```

```
# This code also fixes the size of our input matrices
to 132 x 132
# as graphs of different sizes will result in variable
sizes of
# adjacency matrices.

for line in data:
    splitline = line.split("\t")
    smiles = splitline[0]
    mol = MolFromSmiles(smiles)
    A = ac.GetAdjacencyMatrix(mol)
    x = np.zeros((132, 132))
    x[:A.shape[0], :A.shape[1]] = A
    X.append(x)
    y.append(int(splitline[-1]))

# Reshape array for feeding to tensorflow graph
X = np.array(X)
X = X.reshape((X.shape[0], X.shape[1], X.shape[2], 1))

# Split into training and validation data
X_train, X_test, y_train, y_test = train_test_split
(X, y, test_size=0.20)

# Build model
model = models.Sequential()
model.add(layers.Conv2D(64, (3, 3),
activation='relu', input_shape=(None, None, 1),
padding='SAME'))
model.add(layers.MaxPooling2D((2, 2)))
model.add(layers.Conv2D(32, (3, 3),
activation='relu'))
model.add(layers.GlobalMaxPooling2D())
model.add(layers.Dense(2, activation='softmax'))

model.compile(optimizer='adam',
              loss='sparse_categorical_crossentropy',
              metrics=['accuracy'])
# Fit model
model.fit(X_train, y_train, epochs=10,
          validation_data=(X_test, y_test))
```

As we can see from the example code, in this case we are having to ensure that each input is of the same size (in this case 132×132). Due to our use of the GlobalMaxPooling2D() layer, this is not strictly necessary – as our network is comprised of convolutions up to this layer, it is capable of handling variable sized input. In this case, it is our use of NumPy arrays for the input that means we need to have fixed input sizes. This can be avoided by using more sophisticated data pipelines in TensorFlow via its Data API. Alternatively, inputs can be fed one at a time to the network during training – this is equivalent to setting the mini-batch size to 1.

When running the code, we see that our simple network manages to achieve an accuracy of around 90%. This demonstrates the importance of graph structure for our task – clearly graph structure is very informative for classification of molecule toxicity, making GCN's a good choice. In this case we used only the adjacency matrix, but typically a set of features are also used. These features tell us something about the individual nodes in the graph – for example, our features could be atom types of our nodes. Incorporating this additional information should make our classifier even more powerful, as it will have a comprehensive picture of the graph's structure and content.

6.4 Real World Implementations

While many real-world applications of CNNs are in image processing, they are also incredibly powerful models for a range of other applications. Here we provide links to a range of sophisticated CNN architectures – use these to gain familiarity with the models and consider how you may want to use them in your own work:

1) YOLO – You Only Look Once: a popular and highly effective CNN architecture for computer vision tasks. The idea behind the YOLO architecture is to incorporate object detection (whether an object is in an image) and localisation (where the object is in an image) with a single neural network. Link: https://modelzoo.co/model/keras-yolov3
2) MaskRCNN: a combination of recurrent neural network (RNN) and CNN architectures used for object detection and automatic bounding-box generation for computer vision tasks. Link: https://modelzoo.co/model/mask-r-cnn-keras
3) DenseNet: an architecture comprising densely connected CNN layers. While the original paper applies this to image recognition tasks, this has proven to be useful in a variety of other applications, including

NMR crystallography, protein contact map prediction, and carbon capture and sequestration. Link: https://modelzoo.co/model/densely-connected-convolutional-networks-2

6.5 Summary

In this chapter, we have explored CNNs and learned that these are powerful models when dealing with spatial data. You should now have a good understanding of the fundamental components of CNNs, including:

- Pooling layers
- Convolution layers
- Using TensorFlow to construct CNNs
- Applying convolutional networks to graphs (GCNs)

While the example GCN here is a good illustration of applying convolutions to graphs, as this method is quickly becoming popular, we encourage you to explore recent literature on GCNs and their use within a range of applications from molecular discovery through to network analysis.

6.6 Papers to Read

ImageNet Classification with Deep Convolutional Neural Networks

– Alex Krizhevsky, *et al.*, Proceedings of International Conference on Neural Information Processing Systems, 2012

Link: https://papers.nips.cc/paper/4824-imagenet-classification-with-deep-convolutional-neural-networks.pdf

Abstract: We trained a large, deep convolutional neural network to classify the 1.2 million high-resolution images in the ImageNet LSVRC-2010 contest into the 1000 different classes. On the test data, we achieved top-1 and top-5 error rates of 37.5% and 17.0% which is considerably better than the previous state-of-the-art. The neural network, which has 60 million parameters and 650,000 neurons, consists of five convolutional layers, some of which are followed by max-pooling layers, and three fully-connected layers with a final 1000-way softmax. To make training faster, we used non-saturating neurons and a very efficient GPU implementation of the convolution operation. To reduce overfitting in the fully-connected layers we employed a recently-developed regularization method called "dropout" that proved to be very effective. We also entered a variant of this model in the ILSVRC-2012 competition and achieved a winning top-5 test error rate of 15.3%, compared to 26.2% achieved by the second-best entry.

Notes: One of the landmark papers on CNNs, and the first to demonstrate how deep CNN architectures can be used to achieve state of the art object recognition – learning to successfully classify thousands of objects from a dataset comprising millions of images. The paper demonstrates how the significant learning capacity of the model and ability to learn hierarchical representations of images allows CNNs to achieve significant improvements in classification error rate when applied to the extremely challenging ImageNet dataset. This is also one of the first papers to describe the use of GPUs for accelerating the convolution operations of CNNs.

Convolutional Networks on Graphs for Learning Molecular Fingerprints

– David K. Duvenaud, *et al.*, Proceedings of International Conference on Neural Information Processing Systems, 2015

Link: https://papers.nips.cc/paper/5954-convolutional-networks-on-graphs-for-learning-molecular-fingerprints

Abstract: We introduce a convolutional neural network that operates directly on graphs. These networks allow end-to-end learning of prediction pipelines whose inputs are graphs of arbitrary size and shape. The architecture we present generalizes standard molecular feature extraction methods based on circular fingerprints. We show that these data-driven features are more interpretable, and have better predictive performance on a variety of tasks.

Notes: In one of the key papers on the use of convolutional networks within the physical sciences, Duvenaud et al. demonstrate how GCNs can be used to comprehensively learn prediction pipelines and produce molecular fingerprints with high predictive performance. They demonstrate GCNs' advantages in being able to handle graphical inputs of arbitrary structure, and highlight the additional interpretability afforded by their method.

Deep Residual Learning for Image Recognition

– Kaiming He *et al.*, Proceedings of the International Conference on Computer Vision and Pattern Recognition, 2016

Link: https://openaccess.thecvf.com/content_cvpr_2016/html/He_Deep_Residual_Learning_CVPR_2016_paper.html

Abstract: Deeper neural networks are more difficult to train. We present a residual learning framework to ease the training of networks that are substantially deeper than those used previously. We explicitly reformulate the layers as learning residual functions with reference to the layer inputs, instead of learning unreferenced functions. We provide comprehensive empirical evidence showing that these residual networks are easier to optimize, and can gain accuracy from considerably increased depth. On the ImageNet dataset we evaluate residual nets with a depth of up to 152 layers---8x deeper than VGG nets but still having lower complexity. An ensemble of these residual nets achieves 3.57% error on the ImageNet test set. This result won the 1st place on the ILSVRC 2015 classification task. We also present analysis on CIFAR-10 with 100 and 1000 layers. The depth of representations is of central importance for many visual recognition tasks. Solely due to our extremely deep representations, we obtain a 28% relative improvement on the COCO object detection dataset. Deep residual nets are foundations of our submissions to ILSVRC & COCO 2015 competitions, where we also won the 1st places on the tasks of ImageNet detection, ImageNet localization, COCO detection, and COCO segmentation.

Notes: This landmark paper on image recognition demonstrates how residual connections within deep neural networks can be used to greatly improve optimization of very deep neural networks. In turn, this method has allowed for the development of deep network architectures capable of richer feature embedding. While this paper demonstrates the significant impact of residual connections in CNNs for image recognition, the technique can be used to enhance any very deep network architecture.

7

Auto-Encoders

7.1 Introduction

So far, we have looked at tasks where we have a set of inputs and a set of targets, with the aim being to discover models which will allow us to accurately predict the values of those targets for as yet unseen inputs. This is clearly a very useful and powerful task, but it is far from being the only way we can use deep learning to improve and enrich our research. Consider, for a moment, the case where we are building a model where the target we are looking to try to reproduce *is the input itself*. These kinds of model are known as auto-encoders. What possible use do I have for this kind of model, you may ask? This chapter will demonstrate the power of auto-encoders, including use for non-linear dimensionality reduction, data denoising, and finally some computational creativity!

7.1.1 Auto-Encoders for Dimensionality Reduction

The most basic types of auto-encoder are based on vanilla multilayer perceptrons, and can be thought of as data compressors, where the input and output layers are the same size, and we are minimising a reconstruction error. Essentially, we are trying to learn an identity function. Whilst this may sound trivial, there is one small catch – the number of neurons in the layer between the input and output layers is less than the number of neurons in the input and output layers themselves. If every feature was completely independent of each other, then this would be a very hard task indeed. If, however, there is some redundancy in the representation (that is to say that the data are compressible) then the hidden layer should be able to capture this in much the same way that a principal component analysis (PCA) would.

Deep Learning for Physical Scientists: Accelerating Research with Machine Learning, First Edition. Edward O. Pyzer-Knapp and Matthew Benatan.

Output layer   Hidden layer   Output layer

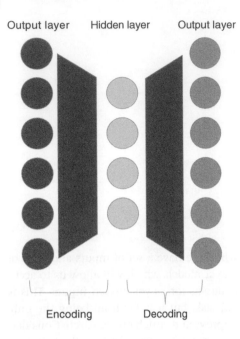

Encoding   Decoding

Figure 7.1 A schematic of a shallow auto-encoder.

This is shown in Figure 7.1, where it can be easily seen that the hidden layer is acting as a compressed representation of the data. The two sets of weights and biases that are learned when training represent functions which encode our input data into the compressed representation and decode our compressed representation back into input space.

So how do we train a model such as this? The easiest way to measure a loss function here is simply the reconstruction error – that is the Euclidean distance between the input and its reconstruction after decoding.

```
import tensorflow as tf

# Set up some constants describing the problem
input_size = 100 # Number of features in the data
compressed_size = 50 # number of neurons in the hidden
layer, here we are compressing data into 50 dimensions
n_inputs = 150 # number of training inputs to the
network (NB this will change per problem, 150 is
arbitrary here)
learning_rate = 0.1 # learning rate
```

```
X = tf.placeholder(tf.float32,[None, n_inputs])
compressed_layer = tf.layers.dense(X,compressed_size)
output_layer = tf.layers.dense(compressed_layer,
input_size)

reconstruction_error = tf.reduce_mean(tf.square
(output_layer - X))

optimizer = tf.train.SGD(learning_rate)
train = optimizer.minimize(reconstruction_error)
```

Of course, we can do this with deeper networks, and these will learn more complex compressions. However, as with all models, overfitting can be a particular problem and care must be taken to avoid this at all costs as an autoencoder which overfits is simply a really expensive identity operation which fails on inputs it has not been trained on – completely useless! One way we can imagine avoiding overfitting is to simply restrict the number of neurons in the hidden layers. Remember, however, that deeper networks are capable of capturing richer, more expressive features – so a shallow network might just not be sufficient for your tasks. Fortunately, if you want to use a deeper network, there are some other tricks you can play.

One particularly effective trick is to use a sparsity constraint on the network. This is effectively telling the network that you only want a small number of *active* neurons, but you do not care which neurons are active. This gives the network the freedom to explore complex compressions, whilst avoiding the risk of overfitting. One of the best ways to achieve this is to use a measure known as the Kullback–Leibler divergence (KL-divergence) between the target sparsity and the actual sparsity. We can write this as:

$$D_{KL} = p \log \frac{p}{q} + (1-p) \log \frac{1-p}{1-q} \tag{7.1}$$

Where D_{KL} is the KL divergence of a target sparsity. Here, q is the measured sparsity and p is the target sparsity.

We measure the sparsity of the network by measuring the average activation of the hidden layer, for each neuron:

```
av_activation = tf.reduce_mean(compressed_layer,
axis=0)
```

Once we have calculated the KL-divergence, we add it as an additional penalty term for the learning. We may want to weigh the contribution, or we can end up with a network which obeys the sparsity rule, but does not learn the structure of the data. We can manage this through weighting the contribution. This is very easy to implement in TensorFlow – let us add this to the simple auto-encoder we created earlier:

```
import tensorflow as tf

# Set up some constants describing the problem
input_size = 100 # Number of features in the data
compressed_size = 50 # number of neurons in the hidden
layer, here we are compressing data into 50 dimensions
n_inputs = 150 # number of training inputs to the
network (NB this will change per problem, 150 is
arbitrary here)
learning_rate = 0.1 # learning rate
target_sparsity = 0.1
sparsity_contribution = 0.2

def kl_d(target_sparsity, measured_sparsity):
    return target_sparsity * tf.log(target_sparsity /
measured_sparsity) + (1 - target_sparsity) * tf.log
((1-target_sparsity) / (1-measured_sparsity))

X = tf.placeholder(tf.float32, [None, n_inputs])
compressed_layer = tf.layers.dense(X,compressed_size)
output_layer = tf.layers.dense(compressed_layer,
input_size)

reconstruction_error = tf.reduce_mean(tf.square
(output_layer - X))
av_activation = tf.reduce_mean(compressed_layer,
axis=0)
sparsity_cost = tf.reduce_sum(kl_d(av_activation,
target_sparsity))
total_loss = reconstruction_error + (sparsity_cost *
sparsity_contribution)
optimizer = tf.train.SGD(learning_rate)
train = optimizer.minimize(total_loss)
```

7.2 Getting a Good Start – Stacked Auto-Encoders, Restricted Boltzmann Machines, and Pretraining

Whilst it would be great to simply have a model which we can plug and play, when we are doing work in which the quality of the weights is crucial (such as dimensionality reduction), it makes sense to stack the deck in our favour as much as possible. One way we can achieve this is through a technique known as pretraining. That is to say, instead of randomly initialising the weights (or using a scheme such as He or Xavier initialisation). Geoff Hinton in his paper in 2006, suggested that a type of model known as a restricted Boltzmann machine (RBM) could be used for this purpose.

7.2.1 Restricted Boltzmann Machines

RBMs – as their name suggests – are a special form of a type of model known as a Boltzmann machine (BM). A BM is a type of neural network based upon the concept of a stochastic neuron. That is to say, in contrast to the neurons you have been used to whose activation is somewhere between 0 and 1, in a BM, neurons will either fire a 0 or 1, but with some probability. This can be written as:

$$P\left(s_i^{t+1} = 1\right) = \sigma\left(\sum_{j=1}^{N} w_{i,j}s_j + b_i\right) \qquad (7.2)$$

Where s_i^t is the output of the neuron at time t (a 0 or a 1), $w_{i,j}$ is the weight of the connection between neurons i and j, b_i is the bias of neuron i, N is the number of neurons in the network, and σ is the logistic function. Unfortunately, BMs are very hard to train, and will oscillate wildly between local configurations. Thankfully, this is not the case with RBMs.

An RBM is a simpler form of BM, and can be thought of as a neural network in which there are only connections between the input layer and a hidden layer. It is trained using a technique known as contrastive divergence (CD-k, where k is the number of CD steps performed). When training an RBM, you sample the hidden layer of the RBM using Gibbs sampling and then you need to find an update matrix to update the weights between the hidden and visible layers. Equation (7.2) shows how to write down the probability of a neuron outputting a 1, and this can be applied to neurons in the hidden and visible layers:

$$P\left(h_j = 1 \mid v\right) = \sigma\left(\sum_{i=1}^{N} w_{i,j} v_i + b_{\mathrm{h}}\right)$$

$$P\left(v_j = 1 \mid h\right) = \sigma\left(\sum_{i=1}^{N} w_{i,j} h_i + b_{\mathrm{v}}\right) \tag{7.3}$$

CD creates a weight update matrix as the difference between the outer products of the probabilities of the hidden layer, given a particular input, or set of inputs.

$$W = v_0 \otimes P(h_0 \mid v_0) - v_k \otimes P(h_k \mid v_k) \tag{7.4}$$

This weight update matrix is then used to update the weights, improving the RBM. In practise, CD-1 (i.e. a chain length of one) has proven to be very effective.

Here is how you might construct an RBM with CD in TensorFlow:

```
import tensorflow as tf

class RBM:

    def __init__(self, input_dim=784, hidden_dim=500,
n_steps_cd=1, momentum=False):
        self.n_visible = input_dim
        self.n_hidden = hidden_dim
        self.k = n_steps_cd
        self.lr = tf.placeholder(tf.float32)
        if momentum:
            self.momentum = tf.placeholder(tf.float32)
        else:
            self.momentum = 0.0
        self.w = weight([input_dim, hidden_dim], 'w')
        self.hidden_bias = bias([hidden_dim], 'hb')
        self.visible_bias = bias([input_dim], 'vb')

        self.weight_velocity = tf.Variable(tf.zeros
([input_dim, hidden_dim]), dtype=tf.float32)
        self.hidden_bias_velocity = tf.Variable(tf.
zeros([hidden_dim]), dtype=tf.float32)
        self.visible_bias_velocity = tf.Variable(tf.
zeros([input_dim]), dtype=tf.float32)
```

```python
    def propagate_up(self, visible):
        pre_sigmoid_activation = tf.matmul(visible,
self.w) + self.hidden_bias
        return tf.nn.sigmoid(pre_sigmoid_activation)

    def propagate_down(self, hidden):
        pre_sigmoid_activation = tf.matmul(hidden, tf.
transpose(self.w)) + self.visible_bias
        return tf.nn.sigmoid(pre_sigmoid_activation)

    def sample_h_given_v(self, v_sample):
        hidden_state = self.propagate_up(v_sample)
        hidden_sample = tf.nn.relu(tf.sign
(hidden_state - tf.random_uniform(tf.shape
(hidden_state))))
        return hidden_sample

    def sample_v_given_h(self, h_sample):
        v_props = self.propagate_down(h_sample)
        v_sample = tf.nn.relu(tf.sign(v_props - tf.
random_uniform(tf.shape(v_props))))
        return v_sample

    def perform_contrastive_divergence(self,
visible_samples):
        # k steps gibbs sampling
        h_samples = self.sample_h_given_v
(visible_samples)
        for i in range(self.k):
            visible_samples = self.sample_v_given_h
(h_samples)
            h_samples = self.sample_h_given_v
(visible_samples)

        h0_props = self.propagate_up(visibles)
        w_positive_grad = tf.matmul(tf.transpose
(visibles), h0_props)
        w_negative_grad = tf.matmul(tf.transpose
(visible_samples), h_samples)
        w_grad = (w_positive_grad - w_negative_grad) /
```

```
tf.to_float(tf.shape(visibles)[0])
    hb_grad = tf.reduce_mean(h0_props - h_samples, 0)
    vb_grad = tf.reduce_mean(visibles -
visible_samples, 0)
    return w_grad, hb_grad, vb_grad


  def train(self, inputs):
    w_grad, hb_grad, vb_grad = self.
perform_contrastive_divergence(inputs)
    # compute new velocities
    new_w_v = self.momentum * self.weight_velocity +
self.lr * w_grad
    new_hb_v = self.momentum * self.
hidden_bias_velocity + self.lr * hb_grad
    new_vb_v = self.momentum * self.
visible_bias_velocity + self.lr * vb_grad
    # update parameters
    update_w = tf.assign(self.w, self.w + new_w_v)
    update_hb = tf.assign(self.hidden_bias, self.
hidden_bias + new_hb_v)
    update_vb = tf.assign(self.visible_bias, self.
visible_bias + new_vb_v)
    # update velocities
    update_w_v = tf.assign(self.weight_velocity,
new_w_v)
    update_hb_v = tf.assign(self.
hidden_bias_velocity, new_hb_v)
    update_vb_v = tf.assign(self.
visible_bias_velocity, new_vb_v)

    return [update_w, update_hb, update_vb,
update_w_v, update_hb_v, update_vb_v]
```

7.2.2 Stacking Restricted Boltzmann Machines

This is all well and good, but how does it relate to pretraining? Imagine for a second that instead of an abstract notion of hidden and visible layers, that

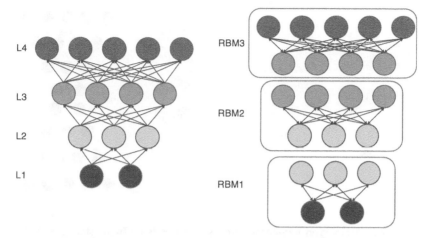

Figure 7.2 Representing a neural network as a stack of RBMs for pretraining.

the visible layer is a layer in a neural network and the hidden layer is the layer directly after it. We can see this graphically in Figure 7.2.

In Figure 7.2, I have coloured each layer a different colour so you can see how the layers in the RBMs relate to the feedforward network. By starting from the first layers in a network we can use the first layer as the visible layer of an RBM with the number of neurons in the hidden layer determined by the number of neurons in the second layer. Once this RBM has been trained with CD-1, we use the hidden layer as the visible layer to a second RBM, and so on and so forth until we have covered every layer in the neural network. When you look at Figure 7.2, you can see that there are the same number of weight matrices in both approaches, thus we can simply transfer the set of weights learned by the stacked RBMs to the feedforward network – a process called "unrolling" (note that this is distinct from unrolling in recurrent networks). Thus, this method provides us with a set of weights for the neural network which are a great starting point for training, so all we require is a smaller number of iterations of backpropagation to fine tune the network. It is easy to see how we can take the network learned in Figure 7.2 and apply it to auto-encoders. Since the reconstruction error is implicit in the stacked RBM approach, we first train the encoder using stacked RBMs. We then mirror these weights, to form the paired decoder. Finally, we use backpropagation (maybe with some sparsity penalty) to fine tune the encoder–decoder pair (if we are happy to break the encoder–decoder symmetry) (Figure 7.3).

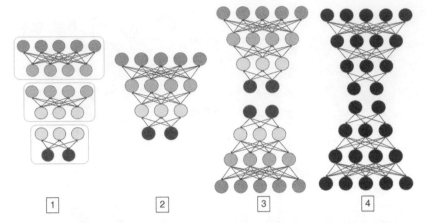

Figure 7.3 Training an auto-encoder from stacked RBMs. (1) Train a stack of RBMs to reproduce an encoder (here, five features to two). (2) Unroll RBMs to create a feed-forward encoder. (3) Create a decoder layer by flipping the weights from the encoder (this is an auto-encoder with tied weights). (4) Optional: fine-tune with back propagation; NB this may break encoder–decoder symmetry.

7.3 Denoising Auto-Encoders

When training an auto-encoder, we have seen how important it is to avoid simply learning an identity transformation. In the previous section, we saw how we can use sparsity to achieve this, but there are of course other methods. One of these is known as the denoising auto-encoder. This is built off the simple premise that by adding some noise to the inputs, and then getting them to reconstruct a clean version of the inputs, we are not mapping like to like, and thus should not have to worry about learning an identity operation.

One common application of this is image data, to which random, Gaussian, noise is added to the images which are input to the network, and then the network is trained to reconstruct the "clean" original images.

Of course, in your day to day work, you are probably presented with noisy data to clean all the time. If you are convinced that all the noise originates from the same place (or is at least all from a similar distribution) then you can imagine that a denoising auto-encoder might be a very useful addition to your arsenal. Simply clean a subset of the data manually, and then train the auto-encoder to replicate your cleaning process. This can speed up workflows significantly. Even better, we can use all the tricks we have learned thus far to improve performance.

7.4 Variational Auto-Encoders

Variational auto-encoders, or VAEs, provide a method of generative modelling based on auto-encoders. Unlike the dimensionality reduction or de-noising applications of auto-encoders discussed so far, VAEs are particularly useful for learning distributions of data, and generating data that fit these distributions. This makes them powerful tools for synthetic data generation or for exploring variations within your data.

The key difference between a VAE and a standard auto-encoder is that the VAE encoding takes the form of a distribution over the latent space, rather than a point estimate. This allows for sampling over this distribution, and therefore the generation of output data that is novel while matching the distribution (Figure 7.4).

To simplify the encoded representation, the distribution of the encoding is chosen to be Gaussian, with a mean and a variance associated with each weight in the encoding, resulting in each weight being represented by the normal distribution $N(\mu, \sigma^2)$. These means and variances are then used to sample from the distribution, producing the latent representation z that is then fed through the decoder to produce the reconstructed input.

To ensure that the VAE converges on the correct distribution, the KL divergence is incorporated into the objective function. The KL divergence will be large if the means (μ) and variances (σ^2) do not match the target distribution, and small if they are close to the target distribution. We also want to ensure that we find the parameters that assign the greatest likelihood to

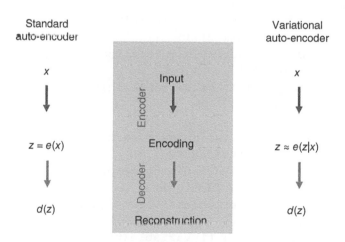

Figure 7.4 Comparison of standard auto-encoder and variational auto-encoder.

our dataset, which we do through incorporating the likelihood into the objective function. This results in following objective function, with the likelihood on the left, and the KL divergence on the right:

$$L = -\log P(X \mid z) + D_{KL}\Big[Q(z \mid X)\big\|P(z)\Big] \qquad (7.5)$$

Loss function for VAE.where $Q(z \mid X)$ is a normal distribution over z values defined as $N(\mu(X), \sigma^2(X))$, and $P(z)$ is a normal distribution defined as $N(0, 1)$.

Let us take a look at constructing a VAE using TensorFlow:

```
import tensorflow as tf

class VAE():
    def __init__(self, input_dim, hidden_dim_1,
hidden_dim_2, z_dim,
                    batch_size):
        self._batch_size = batch_size
        initializer = tf.contrib.layers.
xavier_initializer()
        self.x = tf.placeholder(tf.float32, [None,
input_dim])
        self.z_dim = z_dim

        # Initialize weights
        self._weights_enc = {
            'hidden_1': tf.Variable(initializer
([input_dim, hidden_dim_1])),
            'hidden_2': tf.Variable(initializer
([hidden_dim_1, hidden_dim_2])),
            'z_mu': tf.Variable(initializer
([hidden_dim_2, z_dim])),
            'z_log_sigma': tf.Variable(initializer
([hidden_dim_2, z_dim]))}
        self._biases_enc = {
            'bias_1': tf.Variable(tf.zeros
([hidden_dim_1], dtype=tf.float32)),
            'bias_2': tf.Variable(tf.zeros
([hidden_dim_2], dtype=tf.float32)),
            'z_mu': tf.Variable(tf.zeros([z_dim],
dtype=tf.float32)),
            'z_log_sigma': tf.Variable(tf.zeros
```

```
([z_dim], dtype=tf.float32))}
        self._weights_dec = {
            'hidden_1': tf.Variable(initializer
([input_dim, hidden_dim_2])),
            'hidden_2': tf.Variable(initializer
([hidden_dim_2, hidden_dim_1])),
            'z_mu': tf.Variable(initializer
([hidden_dim_1, n_input])),
            'z_log_sigma': tf.Variable(initializer
([hidden_dim_1, n_input]))}
        self._biases_dec = {
            'bias_1': tf.Variable(tf.zeros
([hidden_dim_2], dtype=tf.float32)),
            'bias_2': tf.Variable(tf.zeros
([hidden_dim_1], dtype=tf.float32)),
            'z_mu': tf.Variable(tf.zeros([input_dim],
dtype=tf.float32)),
            'z_log_sigma': tf.Variable(tf.zeros
([input_dim], dtype=tf.float32))}
        self._softplus = tf.nn.softplus

        # Create network
        self.z_mu, self.z_log_sigma_2 = self._encoder
(self._weights_enc, self._biases_enc)
        sample = tf.random_normal((self._batch_size,
self.z_dim), 0, 1,
                                    dtype=tf.float32)
        self.z = tf.add(self.z_mu, tf.mul(tf.sqrt(tf.
exp(self.z_log_sigma_2)),
                        sample))
        self.reconstruction = self._decoder(self.
_weights_dec, self._biases_dec)

        # Define our negative log likelihood and KL
divergence
        self.log_p = -tf.reduce_sum(self.x * tf.log(1e-
10 + self.reconstruction)
                        + (1-self.x) * tf.log(1e-10 + 1 -
self.reconstruction), 1)
```

```
        self.d_kl = -0.5 * tf.reduce_sum(1 + self.
z_log_sigma_2
                                        - tf.square
(self.z_mu)

                                        - tf.exp
(self.z_log_sigma_2), 1)

    # Define loss function and initialize optimizer
        self.loss = tf.reduce_mean(self.log_p + self.
d_kl)
        self.optimizer = tf.train.AdamOptimizer().
minimize(self.loss)

    def _encoder(self, weights, biases):
        # Create encoder part of network
        layer_1 = self._softplus(tf.add(tf.matmul
(self.x, weights["hidden_1"]),
                                    biases["bias_1"]))
        layer_2 = self._softplus(tf.add(tf.matmul
(layer_1, weights["hidden_2"]),
                                    biases["bias_2"]))
        z_mu = tf.add(tf.matmul(layer_2, weights
["z_mu"]), biases["z_mu"])
        z_log_sigma_2 = tf.add(tf.matmul(layer_2,
weights["z_log_sigma"]),
                            biases["z_log_sigma"])
        return (z_mu, z_log_sigma_2)

    def _decoder(self, weights, biases):
        # Create decoder part of network
        layer_1 = self._softplus(tf.add(tf.matmul
(self.z, weights["hidden_1"]),
                                biases["bias_1"]))
        layer_2 = self._softplus(tf.add(tf.matmul
(layer_1, weights["hidden_2"]),
                                biases["bias_2"]))
        reconstruction = tf.nn.sigmoid(tf.add(tf.
matmul(layer_2,
```

```
weights["z_mu"]),
                                          biases["z_mu"]))
        return reconstruction

    def fit_minibatch(self, X):
        # Fit minibatch, returning minibatch loss
        opt, loss = self.sess.run((self.optimizer,
self.loss),
                                          feed_dict={self.x: X})
        return loss

    def map_to_z(self, X):
        # Map to latent space
        z_mu = self.sess.run(self.z_mu, feed_dict=
{self.x: X})
        return z_mu

    def generate_data(self, z_mu=None):
        # Generate data either by using a specified
point in latent space (z_mu)
        # or by randomly sampling from the latent space
        if z_mu is None:
            z_mu = np.random.normal(self.z_dim)
        gen_data = self.sess.run(self.reconstruction,
feed_dict={self.z: z_mu})
        return gen_data

    def reconstruct(self, X):
        reconstruction = self.sess.run(self.
reconstruction,
                                          feed_dict={self.x:
X})
        return reconstruction
```

7.5 Sequence to Sequence Learning

Sequence to sequence models are very useful when both your input and output has some kind of sequence or time dependent component. Although these are classically used in tasks such as machine translation, their potential is much greater than that.

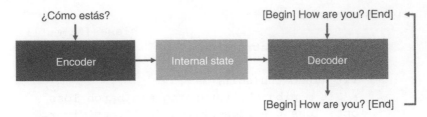

Figure 7.5 Illustration of sequence to sequence model.

As with all auto-encoder-like tasks, sequence to sequence learning is separated into the encoding and decoding functionality. The encoding step takes the source sequence and tries to extract the key information for the input. This is then translated via the decoding step to another sequence which has different properties. An obvious application of this is in machine translation, where we may have an input sentence in Spanish, and an output sentence in English – both sentences contain the same information, but the way in which that information is represented is different. We could also use a sequence to sequence model to learn the relationship between some input numbers and a mathematical relationship between them, e.g. we could train a sequence to sequence model to "learn" to transform the input "[3,7]" into the output "21."

Given their sequential nature, sequence to sequence models typically use recurrent neural networks (RNNs) for both the encoder and decoder. As with any RNN, the encoder RNN will produce both a state and an output. In the case of sequence to sequence modelling, we are only interested in long short-term memory (LSTM's) state, which we pass to the decoder. Using this input as context, the decoder then makes a series of output predictions. For each time t, the decoder takes as input the output from time $t-1$, and makes a prediction conditioned on the LSTM state from the encoder. As illustrated in Figure 7.5, tokens indicating the beginning and end of the sequence are used to tell the model when the sequence starts and finishes.

7.6 The Attention Mechanism

The attention mechanism in neural networks is a method of learning which parts of the input are important for a given output. Just as we may learn to focus our attention on important details in an article or lecture, the attention mechanism provides neural networks with a method of focusing on important data in their input data. This ability to focus on key details and ignore superfluous information makes attention a powerful technique, and it is proven to be key in achieving state-of-the-art results across a range of machine learning problems.

Let us take a typical RNN as an example. In a typical RNN, each prediction relies on the current internal (or hidden) state, but what if we incorporated information from more than just the current hidden state – instead of using information from all hidden states? This is what was proposed in the attention paper by Bahdanau et al. In their paper, they replace the traditional hidden state, h_i, in an RNN with their state vector s_i, which is computed as:

$$s_l = f(s_{i-1}, y_{i-1}, c_i)$$

The important component here is the context vector, c_i, which is defined as a weighted sum of all hidden states $(h_0...h_{T_x})$, from the first hidden state to the state for input at T:

$$c_i = \sum_{j=0}^{T_x} \alpha_{ij} h_j$$

where the weighting coefficient α is defined as:

$$\alpha_{ij} = \frac{\exp(e_{ij})}{\sum_{k=0}^{T_x} \exp(e_{ik})}$$

and e_{ij} is an alignment model given by:

$$e_{ij} = a(s_{i-1}, h_j)$$

This alignment model is fundamental to the mechanics of attention: it scores how well the inputs at position j and outputs at position i match. This, in turn, translates to the weight α_{ij}, which ensures that hidden states relevant to a given output are attributed more weight, i.e. more attention. In doing so, the model is able to incorporate information from all hidden states, but does so with a weighting computed by each state's relevance to the output, rather than naively attributing the same weighting to each hidden state. This allows the decoder to selectively retrieve information from our hidden states, improving efficiency as the decoder does not need to encode information from the entire input.

7.7 Application in Chemistry: Building a Molecular Generator

In this example, we will use what we have learned about auto-encoders to build a network to propose new molecules through using an auto-encoder to learn molecular structures. The example here uses SMILES molecular representations to train an auto-encoder to reproduce these SMILES molecules.

Once trained, we provide synthetic inputs to the latent space that are then processed by our decoder to generate new molecules.

```python
from tensorflow.keras import datasets, layers, models
import tensorflow as tf
from sklearn.model_selection import train_test_split
import numpy as np

from rdkit.Chem import MolFromSmiles
from rdkit.Chem import AllChem as ac

datafile = "nr-er.smiles"
f = open(datafile, "r")
data = f.readlines()
f.close()

feats = []
X = []
y = []

# Get adjacency matrix from each SMILES molecule and
the label
# indicating whether it's toxic (1) or non-toxic (0)

# This code also fixes the size of our input matrices
to 132 x 132
# as graphs of different sizes will result in variable
sizes of
# adjacency matrices.

for line in data:
    try:
        splitline = line.split("\t")
        smiles = splitline[0]
        mol = MolFromSmiles(smiles)
        A = ac.GetAdjacencyMatrix(mol)
        x = np.zeros((132, 132))
        x[:A.shape[0], :A.shape[1]] = A
        X.append(x)
        y.append(int(splitline[-1]))
```

```
        except:
            pass


# Reshape array for feeding to tensorflow graph
X = np.array(X)
X = X.reshape((X.shape[0], X.shape[1], X.shape[2],
1))


# Split into training and validation data
X_train, X_test, y_train, y_test = train_test_split
(X, y, test_size=0.20)


# Create model class for VAE, inheriting from tf.
keras.Model
class VAE(tf.keras.Model):
    def __init__(self, latent_dims):
        super(VAE, self).__init__()
        self.latent_dims = latent_dims
        # Create encoder part of the network
        self.encoder = models.Sequential()
        self.encoder.add(layers.InputLayer
(input_shape=(132,132,1)))
        self.encoder.add(layers.Conv2D(filters=64,
kernel_size=3, strides=(2, 2), activation='relu'))
        self.encoder.add(layers.Conv2D(filters=32,
kernel_size=3, strides=(2, 2), activation='relu'))
        self.encoder.add(layers.Flatten())
        self.encoder.add(layers.Dense(latent_dims*2))


        # Create decoder part of the network
        self.decoder = models.Sequential()

self.decoder.add(layers.InputLayer(input_shape=
(latent_dims,)))
        self.decoder.add(layers.Dense(units=33*33*32,
activation=tf.nn.relu))
        self.decoder.add(layers.Reshape(target_shape=
(33,33,32)))
        self.decoder.add(layers.Conv2DTranspose
(filters=32, kernel_size=3, strides=2,
activation='relu', padding='SAME'))
        self.decoder.add(layers.Conv2DTranspose
```

```python
(filters=64, kernel_size=3, strides=2,
activation='relu', padding='SAME'))
        self.decoder.add(layers.Conv2DTranspose
(filters=1, kernel_size=3, strides=1,
activation='relu', padding='SAME'))


    def gen_sample(self, decode_inti=None):
        if decode_init == None:
            decode_init = tf.random.normal(shape=
(latent_dims*2))
        return self.run_decoder(decode_init,
apply_sig=True)


    def reparam(self, mean, logv):
        rsample = tf.random.normal(shape=mean.shape)
        return rsample * tf.exp(logv * 0.5) + mean


    def run_encoder(self, x):
        mean, logv = tf.split(self.encoder(x),
num_or_size_splits=2, axis=1)
        return mean, logv


    def run_decoder(self, z, apply_sig=False):
        logits = self.decoder(z)
        if apply_sig:
            return tf.sigmoid(logits)
        return logits


    def log_normal_pdf(self, sample, mean, logv,
raxis=1):
        return tf.reduce_sum(-0.5 * ((sample-mean) **
2.0 * tf.exp(-logv) + logv + (tf.math.log(2. * np.
pi))))


    def loss(self, x):
        mean, logv = self.run_encoder(x)
        z = self.reparam(mean, logv)
        x_logits = self.run_decoder(z)
        cross_entropy = tf.nn.
sigmoid_cross_entropy_with_logits(logits=x_logits,
labels=np.float32(x))
```

```
        logpx_z = -tf.reduce_sum(cross_entropy, axis=
[1,2,3])
        logpz = self.log_normal_pdf(z, 0.0, 0.0)
        logqz_x = self.log_normal_pdf(z, mean, logv)
        return -tf.reduce_mean(logpx_z + logpz -
logqz_x)

    def update_grads(self, x, optimizer=tf.keras.
optimizers.Adam()):
        with tf.GradientTape() as tape:
            loss = self.loss(x)
        grads = tape.gradient(loss, self.
trainable_variables)
        optimizer.apply_gradients(zip(grads, self.
trainable_variables))

    def train(self, X_train, epochs, batch_size=10):
        for epoch in range(epochs):
            i = 0
            while i < X_train.shape[0]-batch_size:
                x = X_train[i:i+batch_size].reshape
(batch_size, 132, 132, 1)
                self.update_grads(x)
                i += batch_size

def generate_mols(model, decoder_init):
    return model.gen_sample(decoder_init)


epochs = 100
latent_dim = 100
n_examples = 10
gen_vector = tf.random.normal(shape=[n_examples,
latent_dim])

model = VAE(latent_dim)
model.train(X_train, epochs)

new_mols - generate_mols(model, gen_vector)
```

7.8 Summary

In this chapter, we have explored how we can utilise the seemingly unusual technique of training a model on its own input data to create a potent kind of neural network – an auto-encoder. You should now have a good understanding of:

- Foundations for auto-encoders, including RBMs
- The core components of auto-encoders: encoders and decoders
- Using auto-encoders for dimensionality reduction
- Using auto-encoders for denoising
- Using auto-encoders to generate new data
- Sequence to sequence modelling
- The attention mechanism for neural networks

7.9 Real World Implementations

Several auto-encoder and seq2seq models can be found on ModelZoo. The examples here include an entertaining (and educational) sequence modelling example and a detailed case study of how auto-encoders can be used for latent factor analysis of neuroscience data. Both architectures can be applied within a variety of other contexts – think about how you could make use of one or both of these within your own work:

1) Seq2Seq Chatbot: a fun implementation to play with, and a useful example for learning about sequence to sequence methods. This model uses data from Twitter and the Cornell Movie Dialogue corpus to create a chatbot.
2) LFADS – Latent Factor Analysis via Dynamical Systems: this implements the LFADS sequential variational auto-encoder used for processing time series neuroscience data, but the architecture is useful for working with a wide variety of time series data. Link: https://modelzoo.co/model/lfads

7.10 Papers to Read

Auto-Encoding Variational Bayes

- Diederik P. Kingma, *et al.* Proceedings of the International Conference on Learning Representations, 2014.

Link: https://arxiv.org/abs/1312.6114

Abstract: How can we perform efficient inference and learning in directed probabilistic models, in the presence of continuous latent variables with intractable posterior distributions, and large datasets? We introduce a stochastic variational inference and learning algorithm that scales to large datasets and, under some mild differentiability conditions, even works in the intractable case. Our contributions is two-fold. First, we show that a reparameterization of the variational lower bound yields a lower bound estimator that can be straightforwardly optimized using standard stochastic gradient methods. Second, we show that for i.i.d. datasets with continuous latent variables per datapoint, posterior inference can be made especially efficient by fitting an approximate inference model (also called a recognition model) to the intractable posterior using the proposed lower bound estimator. Theoretical advantages are reflected in experimental results.

Notes: This is the paper that introduced Variational Autoencoders, which have since gone on to be hugely popular for a range of tasks, including feature embedding, denoising, and visualisation.

Sequence to Sequence Learning with Neural Networks

- Ilya Sutskever *et al.* in Proceedings of the Advances in Neural Information Processing Systems conference, 2014.

Link: http://papers.nips.cc/paper/5346-sequence-to-sequence-learning-with-neural-

Abstract: Deep Neural Networks (DNNs) are powerful models that have achieved excellent performance on difficult learning tasks. Although DNNs work well whenever large labeled training sets are available, they cannot be used to map sequences to sequences. In this paper, we present a general end-to-end approach to sequence learning that makes minimal assumptions on the sequence structure. Our method uses a multilayered Long Short-Term Memory (LSTM) to map the input sequence to a vector of a fixed dimensionality, and then another deep LSTM to decode the target sequence from the vector. Our main result is that on an English to French translation task from the WMT'14 dataset, the translations produced by the LSTM achieve a BLEU score of 34.8 on the entire test set, where the LSTM's BLEU score was penalized on out-of-vocabulary words. Additionally, the LSTM did not have difficulty on long sentences. For comparison, a phrase-based SMT system achieves a BLEU score of 33.3 on the same dataset. When we used the LSTM to rerank the 1000 hypotheses produced by the aforementioned SMT system, its BLEU score increases to 36.5, which is close to the previous best result on this task. The LSTM also learned sensible phrase and sentence representations that are sensitive to word order and are relatively invariant to the active and the passive voice. Finally, we found that reversing the order of the words in all source sentences (but not target sentences) improved the LSTM's performance markedly, because doing so introduced many short term dependencies between the source and the target sentence which made the optimization problem easier.

Notes: This paper introduced the sequence to sequence (or seq2seq) family of neural network models. The paper demonstrates seq2seq's considerable advantages in natural language processing tasks, significantly outperforming the then-state of the art through the innovative use of their newly proposed architecture.

8

Optimising Models Using Bayesian Optimisation

8.1 Introduction

Throughout this book, we have seen how we can use deep learning methods for many different types of task – building accurate models for subtle effects with multilayer perceptrons, image recognition and processing with convolutional networks, and sequence prediction with recurrent networks such as LSTMs. Up until this point, we have always provided the hyperparameters for the architecture (what the network looks like) and the training (how we optimise, or train, the network for our task), but as a newly christened deep learning aficionado, you may be wondering how we came up with them (and how you would come up with them when you use these techniques in real life!). Well, there are many different approaches to this but they generally fall into two categories:

- brute force
- optimisation

There are definite pros and cons to each of the approaches, and in this chapter we will work through some of the most common methods: grid, random, greedy, and a powerful method which may be new to some readers – Bayesian optimisation.

8.2 Defining Our Function

Throughout this chapter, we will look at search strategies for optimising the construction of a model. For clarity, we will define it here as a simple multilayer perceptron (MLP) with scaled inputs and targets where we are

Deep Learning for Physical Scientists: Accelerating Research with Machine Learning, First Edition. Edward O. Pyzer-Knapp and Matthew Benatan.
© 2022 John Wiley & Sons Ltd. Published 2022 by John Wiley & Sons Ltd.

looking to optimise the validation error, altering the dropout, the learning rate, and the number of neurons in each of the three layers.

```
s1 = StandardScaler()
s2 = StandardScaler()

Xt_scaled = s1.fit_transform(X_train)
yt_scaled = s2.fit_transform(y_train)

def eval_model(params):
    with graph.as_default():
        l1, l2, dropout1, dropout2, batchsize = params
        model = Sequential()
        model.add(Dense(int(l1), activation='relu',
input_shape=(X_train.shape[1],)))
        model.add(Dropout(float(dropout1)))
        model.add(Dense(int(l2), activation='relu'))
        model.add(Dropout(float(dropout2)))
        model.add(Dense(1, activation='sigmoid'))
        model.compile(loss='mse',
             optimizer=RMSprop(), metrics=[metrics.
mae])
        history_callback = model.fit(Xt_scaled,
yt_scaled, epochs=50, batch_size=int(batchsize),
verbose=0)
        scores = model.evaluate(Xt_scaled, yt_scaled,
verbose=0)
        return float(scores[1])
```

8.3 Grid and Random Search

"Obvious" methods such as using a grid search, or some variant of a random search fall into the first category; that of brute force. Advantages of methods such as this is their ease of implementation – you should be able to stand up these methods in a matter of hours – and the fact that they parallelise very well. Indeed, since every job in a grid or random search is independent of the result of previous jobs, you can run as many of them at once as your machine allows.

A random search is, as its name suggests, a random selection of points. It thus assumes no structure of the search space, and therefore does not exploit any clever tricks. It is, however, still used by some people for performing searches. Why? The answer is because it is very simple to implement, and it is extremely easy to parallelise, since each sample is independent of all the others. It is also worth pointing out for fairness that because random searches do not assume any structure then they also cannot fall prey to search space misspecification – where an underlying model gets the structure wrong, with catastrophic implications for the search efficiency.

If you have some idea about the structure, but still want to use a random-type search, there is a half-way house method which you may have already come across known as the Latin hypercube. Latin hypercube sampling (LHS) aims to spread the places where it samples more evenly across all possible values by exploiting some of the structure. LHS achieves this by partitioning the distribution of each feature into N intervals of equal probability, and selects one sample from each interval. Correlation is avoided by shuffling the sample for each input.

8.4 Moving Towards an Intelligent Search

If we want to get a bit more intelligent about how we select hyperparameters (which will save us time, and most probably money too!) we can take some of the techniques used in the area of optimisation and apply them to this problem. This is not as straightforward as random or grid searching, as we now need to get more information about the underlying function – typically this involves calculating a gradient to show us the best direction for the optimisation to be moving. Unfortunately, as our networks get deeper and more complex, the number of parameters, and the way they interact with each other, gets more and more complex to calculate. Add to this the fact that finding a good set of hyperparameters is a global optimisation problem, it is easy to see why this can scare people off.

8.5 Exploration and Exploitation

When you are trying to optimise a function, there are two tensions which must be traded off against each other:

- The acquisition of new knowledge (aka exploration)
- The use of current knowledge to drive the optimisation (aka exploitation)

Balancing these two tensions is key; too much exploitation and you will have a search which is overly local, and too much exploration and you will have a global search, but one which is very inefficient. Why is this? Well, in order to understand why, we should think about what is driving these two types of search. The best way to do this is to consider two extreme search strategies; greedy search (purely exploitative) and diversity search (purely exploratory).

8.6 Greedy Search

Greedy searches are a pretty common strategy for data driven optimisations in part due to their simplicity to implement and their reasonable performance on simple problems. They represent the extreme of the exploitative strategy which we discussed earlier. The algorithm is pretty easy to understand, and is shown in Figure 8.1. The basic logical flow is to train a model on the data you have, use this model to rank candidates and then add the best candidates to your training set and retrain your model. Wash, rinse, and repeat – as simple as that. However, you should consider the pitfalls of exploitative search strategies such as greedy search.

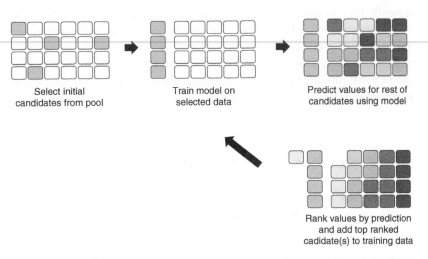

Select initial candidates from pool

Train model on selected data

Predict values for rest of candidates using model

Rank values by prediction and add top ranked cadidate(s) to training data

Figure 8.1 Schematic for greedy search.

8.6.1   Key Fact One – Exploitation Heavy Search is Susceptible to Initial Data Bias

In order to have an exploitative search, we must of course have a model to exploit. This model must be seeded with some initial data points, which are often acquired randomly. In an exploitative search we choose the points which this model predicts to be optimal, add these points to the training set and repeat. Thus, any bias in the initial model due to the selection of the starting training set will be reinforced by each epoch of search, which means that the choice of initial data can subtly determine the direction of your entire search – something which is not desirable at all.

Let us see what a greedy search strategy looks like in code, using a simple linear regression for our model:

```
import numpy as np
from sklearn.linear_model import LinearRegression

epochs = 10
n = X.shape[0]
init_idx = np.random.choice(np.arange(n),
replace=False)

X_train = X_train[init_idx]
y_train = np.array([eval_model(x) for x in X_train])
y_init = y_init.reshape(-1,1)
X_pool = X_init[:]
X_pool = np.delete(X_pool, init_idx, axis=0)

reg = LinearRegression()
reg.fit(X_train, y_train)
for i in range(epochs):

    p = reg.predict(X_pool)
    idx = np.argmax(p)
    X_new = X_pool[idx]
    y_new = eval_model(X_new)
    X_train = np.vstack(X_train, X_new)
    y_train = np.vstack(y_train, y_new)
    X_pool = np.delete(X_pool, idx)

best_value_idx = np.argmax(y_train)
best_value_X = X_train[best_value_idx]
```

When you see this method "in the wild" you will probably not see a purely greedy search used, instead you will see a simple alteration to this, known as "epsilon greedy." Epsilon greedy search tries to avoid the data bias by baking in a mechanism to deviate from greedily following the model predictions. At every epoch a random number is generated, and if this number falls below a critical threshold, a random candidate is chosen to be added to the training set, rather than the candidate selected by the model. The value for this threshold is called "epsilon," thus the name "epsilon greedy." The higher the value of epsilon, the more likely a random structure is to be selected, which allows some sort of control over the overly exploitative tendencies of greedy search. Unfortunately, there is no way to know, a priori, what the correct value for epsilon is, so you are forced to either guess a reasonable value or to tune your optimiser; which may itself be very expensive. Fortunately, epsilon greedy is also very easy to implement – let us take a look, again using linear regression as our model.

```python
import numpy as np
from sklearn.linear_model import LinearRegression

epochs = 10
epsilon = 0.2
n = X.shape[0]
init_idx = np.random.choice(np.arange(n),
replace=False)

X_train = X_train[init_idx]
y_train = np.array([eval_model(x) for x in X_train])
y_init = y_init.reshape(-1,1)
X_pool = X_init[:]
X_pool = np.delete(X_pool, init_idx, axis=0)

reg = LinearRegression()
for i in range(epochs):
 reg.fit(X_train, y_train)
 if np.random.random() < epsilon:
   idx = np.random.randint(X_pool.shape[0])
 else:
   p = reg.predict(X_pool)
   idx = np.argmax(p)
 X_new = X_pool[idx]
 y_new = eval_model(X_new)
```

```
X_train = np.vstack(X_train, X_new)
y_train = np.vstack(y_train, y_new)
X_pool = np.delete(X_pool, idx)

best_value_idx = np.argmax(y_train)
best_value_X = X_train[best_value_idx]
```

8.7   Diversity Search

In some ways, the polar opposite of a greedy search is a "diversity" search. This is sometimes known as maximum entropy searching, as you are maximising the information entropy at each stage of the search.

A diversity driven search is purely exploratory, and at each stage the most "different" candidate is added to the "training" pool. This means that this can be an effective way to search a poorly structured landscape, where there is little to exploit in an optimisation. It is also a good method for building generalizable models quickly, with as small a training set at possible. Diversity search is demonstrated graphically in the below figure.

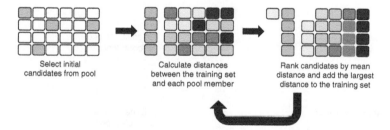

Select initial
candidates from pool

Calculate distances
between the training set
and each pool member

Rank candidates by mean
distance and add the largest
distance to the training set

If you are using a diversity search to screen large spaces thoroughly, it can be very efficient to parallelise this search. As you can see from the above figure, there is no *need* for a model to perform a diversity search, as the distances are only dependent on the features. This means that you can determine which candidates will be tested before you start and try to calculate as many of them in parallel as you have resource for.

Note – using model uncertainty for diversity search. If you do have an underlying model you want to use (maybe because the uncertainty estimates are stronger than a classic distance only metric) it is very simple to adapt this approach. Instead of selecting candidates with the largest mean distance, you simply select the candidate with the highest predictive uncertainty.

8.8 Bayesian Optimisation

Bayesian optimisation is a method which can really help us to be more intelligent when searching for parameters. It is what is known as a "black box" optimiser, which means that it does not need to calculate, or even know about, gradients, and whilst it can appear very complicated, the basics are pretty easy to understand, and will get you a good jump start into creating powerful deep learning models.

Let's start with a high-level overview of what Bayesian optimisation does. In some ways it is a very similar process to that which we use to do our research, you start with some set of opinions, and from that you make predictions on the outcomes of many events, most likely with some degree of uncertainty. You then balance the risks and rewards of collecting some data (maybe it is risky, but could be game changing) collect what you decide to be the most promising candidate, observe the outcome of your experiment, and use that to get a better idea of what is going on, enabling you to design a new experiment. Each of these steps is mirrored in Bayesian optimisation, and the overlap is shown in the below figure:

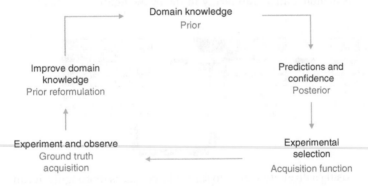

This can seem a little complicated so let us go through this one by one.

8.8.1 Domain Knowledge (or Prior)

In your experiments, you often have an opinion, a reason you chose to perform that experiment. This can be based on past experiences, which you use to assimilate some mental picture of what output results from a certain set of parameters. In Bayesian optimisation this is known as a prior, and we use it to build a model with which we can predict the outcomes of as-yet untried parameters.

The whole of Bayesian optimisation builds from a relatively simple equation – Bayes Rule. It is given as (Figure 8.2):

It is through this equation that we build our experiences into the model – it is $P(A)$.

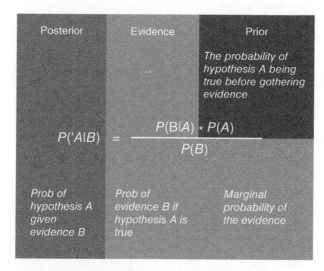

Figure 8.2 Bayes rule.

When we talk about building in our experiences, there are two places in which we can do this. One, obviously, is through supplying the algorithm with the results of events which have taken place – this is known as the evidence. The second is through how the algorithm combines pieces of evidence, using a framework called a kernel. This essentially tells our algorithm how to relate how close two pieces of evidence are. This is all very well and good, but the complexity is starting to stack up! Why do we need to go to all this effort? Let us take a slight detour and go through an example which should demonstrate this point efficiently.

Let's say we have some training data:

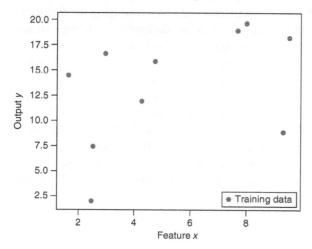

we will initially assume that it is linear since we are all (hopefully!) happy with basic linear regression.

$$Y = \sigma_1 X + \sigma_2$$

(let us ignore for the moment the irreducible error)

What we are trying to do here is to find value for σ_1 and σ_2 to minimise the error between what we predict (using our fitted equation) and what we see (in real life). I am sure we have all done this many times, and here is an example.

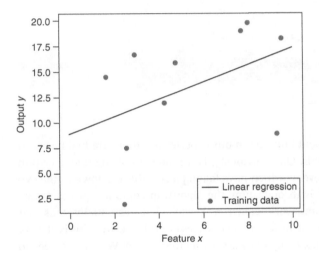

As you can see, this is not idea in a number of ways. Firstly, the data is not clearly correlated with the prediction. Secondly, there is no way of knowing whether this prediction error is due to overfitting, underfitting, or some other artefact of data bias. Can we do better? Well, we can use the Bayes rule to produce a distribution for both of these values, which updates as new pieces of evidence are collected. Since this is a distribution, not a single point, we can get a new piece of information out of our math – the predictive variance. This is very important as we now have a measure of how sure the algorithm is about the predictions it is making. Here is the kind of thing we can expect using a simple Bayesian model known as Bayesian Ridge Regression. You can find this functionality in many packages, including scikit-learn, but as this is only a stopping point on our journey, we will leave further investigation into Bayesian Ridge Regression to the curious reader.

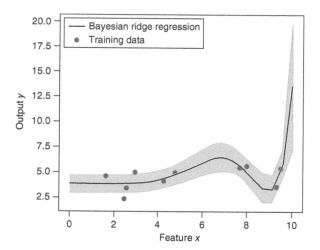

This offers an improvement. Although the model still has problems (for example, see how it veers off to very high values at around $X = 10$), the error is at least somewhat correlated with the uncertainty (represented by the shaded area).

In these kinds of probabilistic models (i.e. models with uncertainty aware predictions), we have two ways of encoding our beliefs – through evidence and through how we relate that evidence. Up until this point, we have treated the relationships between data points through assuming that this data can be fitted using a specific functional form, and a set of coefficients. We could try lots of different functional forms to improve this, but that is very tiresome and prone to bias (see overfitting and underfitting). Is there a way in which we can get away with not specifying the functional form? Welcome to the world of non-parametrics! There are many different ways to do non-parametric regression, but we will spend some time looking at a particular model known as a Gaussian process.

8.8.2 Gaussian Processes

A Gaussian process is a collection of random variables, which has the property that the joint distribution of any finite subset of them is Gaussian. In other words, a process is to a function as a function is to a variable. This is very useful to us, as it gives us a mechanism for describing and evaluating a variable using an infinite number of functions. For a very in-depth mathematical description of GPs, we direct the reader to Carl Rassmussen's excellent book; Gaussian Processes for Machine Learning. Luckily for the eager reader, you do not need to implement GPs yourself from scratch – there are many libraries which will provide you with the tooling to do this. For a basic implementation, as with many machine-learning tasks, scikit-learn offers a good place to start. For more advanced users, two more useful libraries are the excellent GPy which is built from Neil Lawrence's group in Sheffield University, or GPFlow, which originated from Cambridge. GPFlow has the advantage of the fact that it utilises the TensorFlow application programming interface (API) which we have been using throughout this book and so gives us easy access to graphical processing unit (GPU) acceleration when required. When we look at some examples later on, in order to keep things simple and transparent, we will use the Scikit-learn Gaussian Process module, which is very intuitive, and follows the Scikit-learn API, which means that it will feel familiar if you have done any machine learning with this library before.

8.8.3 Kernels

Most of the magic behind a Gaussian process is contained within a construct called the covariance matrix – which is essentially a matrix of measurements between every pair of points. Clearly how we take these measurements is going to greatly affect the model, so it is important to understand how a kernel works.

But first, what is a kernel? Kernels are wonderful constructs which allow us to relate data in such a way that it is possible to construct models which have characteristics we know a priori; that is to say it is a way of infusing expertise into the model. For example, if we know that our data is varying periodically, we can use a kernel with periodic characteristics to relate that knowledge to the model. It is no stretch to say that the kernel defines all the generalisation properties of a Gaussian process. There are a bewildering array of kernels out there to choose from, and indeed you can make your own if you are so inclined. For a kernel to be valid, it must be positively semi-definite (i.e. its eigenvalues are all non-negative). This implies that it is both symmetric, and invertible. Kernels tend to fall into families, and so we will look at a few examples of some of the major classes of kernel:

8.8.3.1 Stationary Kernels
8.8.3.1.1 RBF Kernel One of the most commonly used kernels is the RBF (radial basis function) kernel. You may sometimes see this kernel called the squared exponential kernel, the Gaussian kernel or the exponentiated quadratic kernel. This is because it has several appealing properties – it is universal, and infinitely differentiable (functions from this kernel have

infinitely many derivatives). It is also fairly simple to parameterise, with only two variables contributing to its functional form;

$$k_{SE}(x_p, x_q) = \sigma_f{}^2 \exp\left(-\frac{1}{2l^2}(x_p - x_q)^2\right)$$

The first parameter l is a lengthscale, which determines the length of the "wiggles" in your function. This can be thought of as how sensitive the model is to changes in feature-space. It also gives you a measure of generalisation, as clearly interpolating to a resolution less than l does not have much significant meaning. The second parameter $\sigma_f{}^2$ is a scale function you see in many kernels which is used to describe the average distance of your function away from its mean. You can also think of $(x_p - x_q)^2$ as the squared exponential distance between two points, although it is possible to use any true metric here, and recover a valid kernel.

If you sample from an RBF kernel, you get functions which look like this:

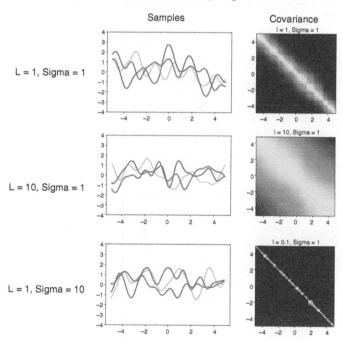

8.8.3.2 Noise Kernel

As you have undoubtedly observed, it is rare that you are presented with perfectly clean data. Gaussian processes are able to handle this kind of data, so long as you make a few assumptions – primarily that the noise is additive, independent and identically distributed (commonly known as iid). Assuming it is also Gaussian, we come to the following statement for a Gaussian process prediction through adding a noise function to the kernel

$$\begin{bmatrix} y \\ f_* \end{bmatrix} \sim \mathcal{N}\left(0, \begin{bmatrix} K(X,X) + \sigma_n^2 I & K(X,X_*) \\ K(X_*,X) & K(X_*,X_*) \end{bmatrix} \right)$$

Where $\sigma_n^2 I$ denotes the noise model.

We can fit the noise level to the data through optimising on the log likelihood. Sometimes, there is the possibility to explain a significant amount of variance from a large noise model. This can be simply tested by comparing the length scales of the model and the noise level (large length scales are indicative of low sensitivity).

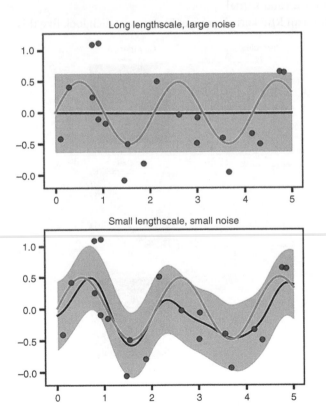

In general, this kind of white noise can help the optimisation of hyperparameters of the GP, and produce more robust predictions.

8.8.4 Combining Gaussian Process Prediction and Optimisation

Well once we have a Bayesian model, we can start to use it to perform optimisation. There are different approaches to optimisation:

1) Improvement-based (I want to find the best solution in the smallest time)
2) Information-based optimisation (I want to build the most general model with the least data)

Let us first think about improvement-based optimisation, as it is something which is very conceptually easy to think about. Clearly the first thing to do is to think about what we mean by improvement. In the greedy case we thought about earlier, improvement was simply how much better than the current best a candidate was predicted to be. Now, though, we have access to more information – we also know how certain the particular model is of its prediction. This allows us to come up with a way of talking about improvement which also takes model certainty into account:

$$\gamma(x) = \frac{\mu(x) - y_{max} + \varepsilon}{\sigma(x)}$$

Where $\gamma(x)$ is known as the *improvement* for a candidate with features x, $\mu(x)$ is the predicted y value for the candidate, y_{max} is the best y value observed thus far, and $\sigma(x)$ is the variance of the prediction.

Let us see how we would express that in code:

```
def improvement(y_pred, y_best, sigma, epsilon):
    imp = (y_pred - y_best + epsilon) / sigma
    return imp
```

Once you have a way of expressing improvement, building up a strategy for using this value becomes much easier. These strategies are known as acquisition functions, and there are a few common ones which are worth thinking about.

8.8.4.1 Probability of Improvement

Perhaps the simplest acquisition function is probability of improvement, often known as PI. It simply asks the question – which candidate has the highest probability of improving over my current optimum?

It is formulated as follows:

$$\alpha_{PI} = \Phi(\gamma(x))$$

Where Φ represents the cumulative distribution function (CDF) of a normal distribution.

Let us see how we would calculate that value, using Python.

```
from scipy.stats import norm

def prob_imp(y_pred, y_best, sigma, epsilon):
    imp = improvement(y_pred, y_best, sigma, epsilon)
    p_i = norm.cdf(imp)
    return p_i
```

Whilst it is very simple to understand and implement, PI is not used widely any more. This is because it has been shown to favour overly local searches. To understand why, we should look back at how our Bayesian model works. Under a Gaussian process, things which are similar are predicted to behave similarly, and the uncertainty is derived from how dissimilar two points are. Thus, points which are predicted to have a high PI are likely to be similar to the current optimum, which drives the local search.

8.8.4.2 Expected Improvement

One way to improve the performance of the acquisition function is not just to consider the PI, but also the magnitude of that improvement. This is achieved using the "expected improvement (EI)" acquisition function, which is written down as:

$$\alpha_{EI} = \sigma(x)(\gamma(x)\Phi(\gamma(x)) + \varphi(y(x)))$$

We can see this in code;

```
from scipy.stats import norm

def exp_imp(y_pred, y_best, sigma, epsilon):
    imp = improvement(y_pred, y_best, sigma, epsilon)
    e_i = exp_imp = sigma * (imp * norm.cdf(imp) + norm.pdf(imp))
    return e_i
```

EI is the most commonly used acquisition function in Bayesian optimisation as it has good guarantees of results, and is relatively straightforward to implement.

8.8.5 Balancing Exploration and Exploitation

For both EI and PI, a common trick to improve performance further is to add a margin to the improvement equation, turning it into:

$$\gamma(x) = \frac{\mu(x) - y_{\max} + \varepsilon}{\sigma(x)}$$

Where ε is a small number which is used to inflate the current "best" candidate, essentially demanding an improvement of at least ε. The larger this margin is, the more the variance of each prediction is explored, leading to more exploration. A common value for ε is 0.1, although this is clearly problem specific.

We have recently proposed a slight twist to this formalism, where the value of ε is determined through investigating the confidence of the model over all the problem space. This ensures that when model confidence is low, we explore (add more knowledge to the model) and when model confidence is high, we exploit (trust the model). We can write down our changes like this:

$$\chi = \frac{\mu(x) - y_{\max} + |C_v|}{\sigma(x)} \quad C_v = \frac{\overline{\sigma(x)}}{y_{\max}}$$

Where χ is our new version of improvement, called contextual improvement.

8.8.6 Upper and Lower Confidence Bound Algorithm

This algorithm is often seen in the related area of multi-arm bandits, and is conceptually very simple. Given a model with reasonable uncertainty estimates, the upper confidence bound (UCB) or lower confidence bound (LCB) algorithms work on a principle of maximal optimism or pessimism respectively. For simplicity, and because it is the more common setting, we will continue to focus solely on the UCB algorithm. The UCB algorithm seeks to sample the largest value plausibly possible. Thus, in order to calculate the UCB acquisition value for any point, you simply use the following equation.

$$\alpha_{\text{UCB}} = \mu(x) + \beta^{\frac{1}{2}} \sigma(x)$$

Where $\beta^{\frac{1}{2}}$ is a value known as the *confidence level*. As can be seen in the below figure, the UCB algorithm hinges on the definition of what a "plausibly large" value for a given prediction is. That is to say, given the prediction and its inherent uncertainty, how large could the value plausibly be in real life. Clearly, this hinges on how confident you are in the quality of your uncertainty value, which is where the confidence level ($\beta^{\frac{1}{2}}$) comes in. The larger the value of $\beta^{\frac{1}{2}}$, the more the algorithm seeks the extreme of the distribution to find "plausibly large" values.

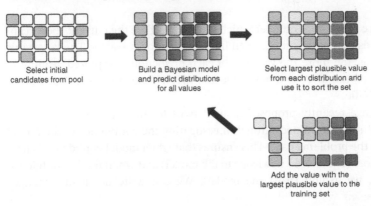

Select initial
candidates from pool

Build a Bayesian model
and predict distributions
for all values

Select largest plausible value
from each distribution and
use it to sort the set

Add the value with the
largest plausible value to the
training set

So why does this work? It is easiest to think of this on an intuitive level. When you take an optimistic action in the way UCB does, it can go one of two ways – either you find a desirable result (the optimism was rewarded) or you don't (the optimism was punished). If the optimism was rewarded, then things have already gone well, but if it was punished then you learn more about distinguishing things which look good (but don't pan out) with things which are actually good. After doing this sufficiently often, you have learnt to distinguish well enough that your optimism is rewarded more than it is punished. Of course, there is no free lunch, and in much the same way we saw the effects of explore/exploit trade-off in EI, the tuning of the confidence level performs the same task in UCB.

8.8.7 Maximum Entropy Sampling

Sometimes, the goal is not a classic optimisation problem, but instead you want to collect data in such a way that you can efficiently build a model using as little data as possible. A common tool for this task is known as maximum entropy sampling (MES).

MES does not consider the target value at all, but instead focuses on the information contained within each data point. One way to think about this is to imagine your problem as a box, which you will consider suitably full when ¾ of the volume is filled. Each data point you collect takes up a volume equal to the amount of information it contains. What MES does is to pick the largest items first, so as to fill up your box as quickly as possible. Of course it is not quite as simple as that, as the information gain is of course relative to the information you have already collected, and so after putting an item in the box, the volumes of the remaining pieces are recalculated. Whilst this may sound like a problem, it actually provides some help to the sampling, as when an item is put into the box, all similar items get smaller so the volume (information) is concentrated in fewer items.

The problem then becomes how to quantify the information gain. Fortunately for us, under a GP prior, we can consider the information gain to be directly related to the predictive uncertainty of a particular point, and thus we can simply rank our points by their uncertainty, and choose the most uncertain to add to the training data. Since the uncertainty is related to the distance, the quick reader will spot the similarity between MES and diversity search.

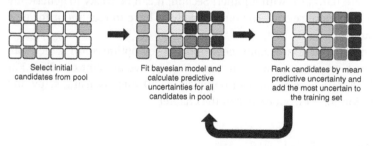

Select initial candidates from pool

Fit bayesian model and calculate predictive uncertainties for all candidates in pool

Rank candidates by mean predictive uncertainty and add the most uncertain to the training set

8.8.8 Optimising the Acquisition Function

Once you have formulated your acquisition function, the next task is to find its maximum. There are two approaches to this, each with their pros and cons. The simplest is to simply enumerate the space you are looking to optimise over, and then find the solution which has the largest acquisition value. This is particularly suitable if your space is naturally discrete, such as a library of molecules. If, however, you are instead discretising a continuous space then this can get very memory hungry as the dimensionality increases. The reason for this can be easily seen by thinking about the number of points in the hypercube which covers increasing numbers of dimensions. Let us say we are evaluating 10 points per dimension and that each point requires 24 bytes (in reality it is more with overheads):

```
1D - 10
2D - 10 * 10 = 10² =  100
```

```
3D - 10 * 10 = 10³ = 1000
4D - 10 * 10 = 10⁴ = 10000
5D - 10 * 10 = 10⁵ = 100,000
6D - 10 * 10 = 10⁶ = 1,000,000
```

Even on this toy problem, there is 24 MB of data just to enumerate the grid of input vectors!

An alternative is to use a second optimiser which allows you to optimise the acquisition function directly (since you know the closed form of the acquisition function, this is possible). This is significantly less heavy on memory, but has a few drawbacks. Firstly, high dimensional optimisation is hard, even using local optimisers, and there is no guarantee that the solution these optimisers find if the true maximum. Indeed you are simply replacing one NP hard task with another! Second, it can be tricky to generically define a constraint setting, so you are likely to have to redefine it for new problems.

Fortunately, there are many good toolkits for optimisation which can allow you to experiment without expending a huge amount of effort. For example, here is a way you can optimise the EI directly using scipy and our function for EI from earlier in the chapter.

```
from scipy import minimise
from sklearn.gaussian_process import
GaussianProcessRegressor as gpr

# Taking X_train and y_train from previous examples

model = gpr.fit(X_train, y_train)

def af_opt(v, ybest=ybest, model=model):
        y_pred, sigma_pred = self.model.predict(v,
return_std=True)
        acq = ei(y_pred, y_best, sigma, epsilon)
        return acq

def opt_af(y_best):
        res = minimise(af_opt, args=y_best, model)
        return res.x
```

You can – and we encourage you to – experiment with different optimisers, each of which allow different settings, and functionalities. As a default, this will be the common Broyden–Fletcher–Goldfarb–Shanno (BFGS) local optimiser.

8.8.9 Cost Sensitive Bayesian Optimisation

Of course everything we have looked at assumes that every data point has the same "cost" to acquisition. In reality, this is rarely the case, and this is especially true in the scientific world. In order to combat this, we can think about optimising in a cost sensitive manner. The most common way that we will achieve this is to warp the acquisition function by the cost to acquire the point. This produces a kind of "value for money" metric which can help guide us in cases where cost is a problem.

If we know how to calculate the cost of something, then this is very easy to do indeed. We simply calculate the acquisition function value, and divide it by the calculated cost and maximise over the result. Let us see how we do this in code, using a simple cost function.

```python
import numpy as np
from scipy.stats import norm
from sklearn.gaussian_process import
GaussianProcessRegressor as gpr

epochs = 10
epsilon = 0.2
n = X.shape[0]
init_idx = np.random.choice(np.arange(n),
replace=False)

X_train = X[init_idx]
y_train = np.array([eval_model(x) for x in X_train])
y_init = y_init.reshape(-1,1)
X_pool = X_init[:]
X_pool = np.delete(X_pool, init_idx, axis=0)

def exp_imp(y_pred, y_best, sigma, epsilon):
    imp = improvement(y_pred, y_best, sigma, epsilon)
    e_i = exp_imp = sigma * (imp * norm.cdf(imp) + norm.
pdf(imp))
    return e_i
def cost(x):
```

```
        return np.sum(x, axis=0)
model = gpr()
for i in range(epochs):
    model.fit(X_train, y_train)
    y_pred, sigma = model.predict(X_pool,
return_std=True)
    y_best = np.min(y_train)
    aq = exp_imp(y_pred, y_best, sigma, epsilon)
    c = cost(X_pool)
    aq = aq/c
    idx = np.argmax(aq)
  X_new = X_pool[idx]
  y_new = eval_model(X_new)
  X_train = np.vstack(X_train, X_new)
  y_train = np.vstack(y_train, y_new)
  X_pool = np.delete(X_pool, idx)

best_value_idx = np.argmax(y_train)
best_value_X = X_train[best_value_idx]
```

Of course, cost is often a very complex thing to know in advance, especially for systems which are novel. In the world of simulations, where cost would be approximated by the time it takes to run on a cluster, we can often make the assumption that similar looking systems should have similar magnitudes of cost. This is the case, for example, with calculating quantum chemical energies, where we know the cost scales with the number of electrons (but we often do not know the prefactor, which is likely machine dependent). In these cases, a good idea is to model the cost with a separate model and use this to predict a cost. In these circumstances, we have found a Gaussian process is a reasonable choice. In this case, the process for calculating the warped acquisition function will look like this:

```
import numpy as np
from sklearn.gaussian_process import
GaussianProcessRegressor as gpr

epochs = 10
epsilon = 0.2
n = X.shape[0]
```

```python
init_idx = np.random.choice(np.arange(n),
replace=False)

def cost(x):
    return np.sum(x, axis=0)

X_train = X[init_idx]
y_train = np.array([eval_model(x) for x in X_train])
y_train = y_train.reshape(-1,1)
cost_train = np.array([cost(x) for x in X_train])
X_pool = X_init[:]
X_pool = np.delete(X_pool, init_idx, axis=0)

def exp_imp(y_pred, y_best, sigma, epsilon):
    imp = improvement(y_pred, y_best, sigma, epsilon)
    e_i = exp_imp = sigma * (imp * norm.cdf(imp) + norm.
pdf(imp))
    return e_i

def build_cost_model(X_train, cost_train):
    gp = gpr()
    gp.fit(X_train, cost_train)
    return gp

def predict_cost(X_pool, model):
    cost_pred = model.predict(X_pool)
    return cost_pred

model = gpr()
for i in range(epochs):
    model.fit(X_train, y_train)
    y_pred, sigma = model.predict(X_pool,
return_std=True)
    y_best = np.min(y_train)
    aq = exp_imp(y_pred, y_best, sigma, epsilon)
    cost_model = build_cost_model(X_train, cost_train)
    c_p = predict_cost(X_pool, cost_model)
    aq = aq/c
    idx = np.arqmax(aq)
 X_new = X_pool[idx]
 y_new = eval_model(X_new)
```

```
c_new = cost(X_new)
X_train = np.vstack(X_train, X_new)
y_train = np.vstack(y_train, y_new)
cost_train = np.vstack(cost_train, c_new)
X_pool = np.delete(X_pool, idx)

best_value_idx = np.argmax(y_train)
best_value_X = X_train[best_value_idx]
```

It is possible to jointly model the cost and the target (that is to say build a multi-output model which can predict both things at once). This is only really useful, however, when you believe that the target property and cost are likely to share a representation, otherwise the resulting model will tend to underfit both properties, and hurt your optimisation.

8.8.10 Constrained Bayesian Optimisation

Constrained optimisation is a well-studied area, and it should be no surprise that it is possible to perform constrained Bayesian optimisation as well. There are a myriad of techniques for doing this, but we will focus on the simplest methods based on similar principles to cost sensitive optimisation to provide the reader with a gentle introduction. We will also consider two situations – where you can write down the constraint as a rule or rules, and when you cannot.

If you can write down the constraint as a set of rules which can be evaluated then you can easily set the acquisition function to zero when these rules are violated. This is as trivial as it sounds if you are calculating the acquisition function over a discrete set of values, but is only slightly more tricky if you are looking to directly optimise the acquisition function.

8.8.11 Parallel Bayesian Optimisation

8.8.11.1 qEI

The simplest way you can think of to make Bayesian optimisation parallel is to simply take more than one sample at each epoch. That is, instead of only selecting the candidate with the highest acquisition function score, we select the candidates with the top N acquisition scores. When the acquisition function is the common EI function, this is known as qEI. qEI is very easy to implement – simply rank your candidates via acquisition function, and select the top N.

```
import numpy as np
from sklearn.gaussian_process import
GaussianProcessRegressor as gpr

epochs = 10
n_sample = 10
epsilon = 0.2
n = X.shape[0]
init_idx = np.random.choice(np.arange(n),
replace=False)

X_train = X[init_idx]
y_train = np.array([eval_model(x) for x in X_train])
y_init = y_init.reshape(-1,1)
X_pool = X_init[:]
X_pool = np.delete(X_pool, init_idx, axis=0)

def exp_imp(y_pred, y_best, sigma, epsilon):
    imp = improvement(y_pred, y_best, sigma, epsilon)
    e_i = exp_imp = sigma * (imp * norm.cdf(imp) + norm.
pdf(imp))
    return e_i


model = gpr()
for i in range(epochs):
    model.fit(X_train, y_train)
    y_pred, sigma = model.predict(X_pool,
return_std=True)
    y_best = np.min(y_train)
    aq = exp_imp(y_pred, y_best, sigma, epsilon)
    idx = np.argsort(-aq)[:n_sample]
 X_new = X_pool[idx]
 y_new = eval_model(X_new)
 X_train = np.vstack(X_train, X_new)
 y_train = np.vstack(y_train, y_new)
 X_pool = np.delete(X_pool, idx)

best_value_idx = np.argmax(y_train)
best_value_X = X_train[best_value_idx]
```

qEI, as you may have already guessed, is not a particularly high-performing parallel algorithm. Since the GP, and thus the acquisition function, is based on relating data points through a kernel, the acquisition score for each point is not independent of the others. Thus, sampling one point has the potential to massively influence the acquisition score of the others. Better ways to do parallel optimisation include ways to account for this, by either trying to guess at the impact of sampling a point, or to punish sampling similar points. We will look at a few of these methods now.

8.8.11.2 Constant Liar and Kriging Believer

Constant Liar and Kriging Believer are extensions to the qEI methodology which aim to greedily build a batch based upon the synthesis of responses for potential candidates. For the constant liar algorithm, the assumption is that each candidate will return a constant "lie," typically either the highest or lowest value observed to date. Alternatively, the Kriging believer algorithm assumes that the synthetic data will be equal to the predicted mean of the GP model upon which the acquisition function is based. This data is then temporarily used to augment the training dataset, and a new (temporary) GP is fitted and acquisition functions are recalculated to greedily build the batch. When thinking of this in code, we use the same initial pieces of code, and simply change the update between constant liar and Kriging believer.

For both:

```
import numpy as np
from sklearn.gaussian_process import
GaussianProcessRegressor as gpr
```

```
epochs = 10
n_sample = 10
epsilon = 0.2
n = X.shape[0]
init_idx = np.random.choice(np.arange(n),
replace=False)

X_train = X[init_idx]
y_train = np.array([eval_model(x) for x in X_train])
y_init = y_init.reshape(-1,1)
X_pool = X_init[:]
X_pool = np.delete(X_pool, init_idx, axis=0)
```

```
def exp_imp(y_pred, y_best, sigma, epsilon):
    imp = improvement(y_pred, y_best, sigma, epsilon)
    e_i = exp_imp = sigma * (imp * norm.cdf(imp) + norm.
pdf(imp))
    return e_i

model = gpr()
candidates_to_test = []
```

For Constant Liar (max):

```
for i in range(n_epochs):
  gpr.fit(X_train, y_train)
  yp, sp = gp1.predict(X_pool)
  ybest = np.max(y_train)
  nsample = 10
  eis = ei(yp, sp, ybest)
  idx = np.argsort(-eis)[0]
  candidates_to_test.append(X_pool[idx])
  X_train.append(X_pool[idx])
  y_train.append(np.max(y_train))
  np.delete(X_pool, idx)
```

For Kriging Believer:

```
for i in range(n_epochs):
  gpr.fit(X_train, y_train)
  yp, sp = gp1.predict(X_pool)
  ybest = np.max(y_train)
  nsample = 10
  eis = ei(yp, sp, ybest)
  idx = np.argsort(-eis)[0]
  candidates_to_test.append(X_pool[idx])
  X_train.append(X_pool[idx])
  y_train.append(yp[idx])
  np.delete(X_pool, idx)
```

There are more sophisticated ways of doing parallel Bayesian optimisation, but the code for these can get very complicated. Therefore, for clarity, we will discuss the algorithm, and provide a pointer to the paper which introduces the algorithm. It is a good exercise for the reader to try to implement these methods themselves, as being able to take an algorithm from a paper and implement it on your own problems is the hallmark of a strong data-driven researcher.

8.8.11.3 Local Penalisation

Link to paper: http://proceedings.mlr.press/v51/gonzalez16a.pdf

As we have discussed, when we are using a Gaussian Process, we anticipate that target values of candidates which are close in sample space are expected to be highly correlated. Thus, we can expect that a sensible thing to do is to apply a penalty to selecting similar points. The local penalisation method does this by successively penalising the acquisition function around already evaluated points, using a penalisation radius based on the estimated Lipschitz constant of the acquisition function surface. This is achieved in exactly the same manner as a cost aware search, except here, a high cost is equivalent to a high Lipschitz penalty term.

8.8.11.4 Parallel Thompson Sampling

Link to paper: http://proceedings.mlr.press/v70/hernandez-lobato17a.html

Parallel Thompson sampling takes a slightly different approach to building batches. It is based on the realisation that a Bayesian model is an infinite collection of deterministic models. Thompson sampling samples this model space to build a collection of deterministic models, which can act as experts. Each model then ranks the candidates and returns their top candidates. Where the models have seen a lot of data, they are likely to agree and so this will be akin to exploitation. In unexplored areas, however, they are likely to differ, and this is akin to exploration.

8.8.11.5 *K*-Means Batch Bayesian Optimisation

Link to paper: https://arxiv.org/abs/1806.01159

K-means batch Bayesian optimisation (KMBBO for short) was derived as an efficient parallel sampling algorithm when you know the number of "slots" you want to fill (i.e. the size of the parallelism). It works by using *K*-means (where *K* is the number of slots) to divide the acquisition function (for example, EI). The reason this works well is that this unsupervised division of the acquisition function will preferentially peak pick, but if there are more slots than peaks, the spherical repulsion term within the *K*-means algorithm ensures that the remaining slots are filled with highly diverse candidates – and as we have seen before diversity can be very efficient in building good models, which then leads to better formulated acquisition functions in subsequent rounds.

8.9   Summary

By the end of this chapter, you have learnt how to optimise the construction of your deep learning models using Bayesian optimisation. You should now understand:

- The importance of predictive variance in machine learning models
- What a Gaussian process is, and how to make predictions with it
- How to calculate the improvement of a candidate using a GP predictive posterior
- How to use the calculated improvement to determine an acquisition function value, and use that value to optimise your model
- How to extend the acquisition functions to include cost, and constraints
- How to extend the optimisation to sample multiple points at the same time.

8.10   Papers to Read

A Tutorial on Bayesian Optimization of Expensive Cost Functions, with Application to Active User Modeling and Hierarchical Reinforcement Learning

- Eric Brochu, Vlad M. Cora, Nando de Freitas

Link: https://arxiv.org/abs/1012.2599

Abstract: We present a tutorial on Bayesian optimization, a method of finding the maximum of expensive cost functions. Bayesian optimization employs the Bayesian technique of setting a prior over the objective function and combining it with evidence to get a posterior function. This permits a utility-based selection of the next observation to make on the objective function, which must take into account both exploration (sampling from areas of high uncertainty) and exploitation (sampling areas likely to offer improvement over the current best observation). We also present two detailed extensions of Bayesian optimization, with experiments---active user modelling with preferences, and hierarchical reinforcement learning---and a discussion of the pros and cons of Bayesian optimization based on our experiences.

Notes: This is a great tutorial on Bayesian optimization which delves deeper into some of the maths behind this powerful method. There is a detailed discussion on how to select covariance functions and acquisition functions, and it introduces the concept of explore-exploit trade off.

Practical Bayesian optimization of machine learning algorithms

- Jasper Snoek, Hugo Larochelle, Hugo Larochelle, Ryan Prescott Adams

Link: https://papers.nips.cc/paper/4522-practical-bayesian-optimization-of-machine-learning-algorithms.pdf

Abstract: The use of machine learning algorithms frequently involves careful tuning of learning parameters and model hyperparameters. Unfortunately, this tuning is often a "black art" requiring expert experience, rules of thumb, or sometimes brute-force search. There is therefore great appeal for automatic approaches that can optimize the performance of any given learning algorithm to the problem at hand. In this work, we consider this problem through the framework of Bayesian optimization, in which a learning algorithm's generalization performance is modeled as a sample from a Gaussian process (GP). We show that certain choices for the nature of the GP, such as the type of kernel and the treatment of its hyperparameters, can play a crucial role in obtaining a good optimizer that can achieve expertlevel performance. We describe new algorithms that take into account the variable cost (duration) of learning algorithm experiments and that can leverage the presence of multiple cores for parallel experimentation. We show that these proposed algorithms improve on previous automatic procedures and can reach or surpass human expert-level optimization for many algorithms including latent Dirichlet allocation, structured SVMs and convolutional neural networks.

Notes: This was the paper which brought the use of Bayesian optimization for the tuning of machine learning algorithms into the public eye. The authors delve deep into how to make decisions on kernels, and model hyperparameters can affect the outcome of your tuning, and show that this method can achieve better than human level performance.

Parallel and Distributed Thompson Sampling for Large-scale Accelerated Exploration of Chemical Space

- José Miguel Hernández-Lobato, James Requeima, Edward O. Pyzer-Knapp, Alán Aspuru-Guzik

Link: http://proceedings.mlr.press/v70/hernandez-lobato17a.html

Abstract: Chemical space is so large that brute force searches for new interesting molecules are infeasible. High-throughput virtual screening via computer cluster simulations can speed up the discovery process by collecting very large amounts of data in parallel, e.g., up to hundreds or thousands of parallel measurements. Bayesian optimization (BO) can produce additional acceleration by sequentially identifying the most useful simulations or experiments to be performed next. However, current BO methods cannot scale to the large numbers of parallel measurements and the massive libraries of molecules currently used in high-throughput screening. Here, we propose a scalable solution based on a parallel and distributed implementation of Thompson sampling (PDTS). We show that, in small scale problems, PDTS performs similarly as parallel expected improvement (EI), a batch version of the most widely used BO heuristic. Additionally, in settings where parallel EI does not scale, PDTS outperforms other scalable baselines such as a greedy search, ε-greedy approaches and a random search method. These results show that PDTS is a successful solution for large-scale parallel BO.

Notes: This is one of the first papers using Bayesian optimization for the exploration of chemical space. This paper gives more detail on the Thompson sampling algorithm discussed in this chapter, and benchmarks it for a relevant use case against other search strategies. In this study, the authors use a Bayesian neural network to model the underlying function, demonstrating how the Bayesian optimization methodology can be extended beyond Gaussian processes.

Case Study 1

Solubility Prediction Case Study

Let us try to use what we have learned to create a classifier on a real-world data set – predicting the solubility of a particular molecule based upon a cheminformatics fingerprint. You can download the data set from https://pubs. acs.org/doi/suppl/10.1021/ci034243x/suppl_file/ci034243xsi20040112_ 053635.txt. For the purposes of this case study, I have downloaded it to a file called solubility.csv, which I will then read in with a python package called Pandas. You can find out more about pandas at: https://pandas.pydata.org/

CS 1.1   Step 1 – Import Packages

```
# Data prep packages

import pandas as pd
import numpy as np
from sklearn.model_selection import train_test_split
# Deep learning packages for building the model
from keras.models import Sequential
from keras.layers import Dense, Dropout
from keras.callbacks import EarlyStopping
# Cheminformatics packages for creating the inputs
import rdkit.Chem.AllChem as ac
```

Deep Learning for Physical Scientists: Accelerating Research with Machine Learning,
First Edition. Edward O. Pyzer-Knapp and Matthew Benatan.
© 2022 John Wiley & Sons Ltd. Published 2022 by John Wiley & Sons Ltd.

CS 1.2   Step 2 – Importing the Data

```
data = pd.read_csv('solubility.csv')
```

Pandas can show you the data you imported using the .head() function

```
data.head()
```

	Compound ID	Measured log(solubility : mol/l)	ESOL predicted log(solubility : mol/l)	SMILES
0	1,1,1,2-tetrachloroethane	−2.18	−2.794	ClCC(Cl)(Cl)Cl
1	1,1,1-trichloroethane	−2.00	−2.232	CC(Cl)(Cl)Cl
2	1,1,2,2-tetrachloroethane	−1.74	−2.549	ClC(Cl)C(Cl)Cl
3	1,1,2-trichloroethane	−1.48	−1.961	ClCC(Cl)Cl
4	1,1,2-trichlorotrifluoroethane	−3.04	−3.077	FC(F)(Cl)C(F)(Cl)Cl

CS 1.3   Step 3 – Creating the Inputs

Here we will use the common cheminformatics descriptor, MACCS keys, built using the RDKit package. You can find out more about these at: https://www.rdkit.org/

```
mols = [ac.MolFromSmiles(s) for s in data['SMILES']]
X = np.array([ac.GetMACCSKeysFingerprint(m) for m in mols])
y = data['measured log(solubility:mol/L)'].values
```

CS 1.4   Step 4 – Splitting into Training and Testing

This is simple to do using the scikit-learn train_test_split function.

```
X_train, X_test, y_train, y_test = train_test_split(
    X, y, test_size=0.1, random_state=42)
```

CS 1.5 Step 5 – Defining Our Model

This is where we start to bring in the models we know from our studies. We will create a function to return a model, the reasons for which will become clear later. Let us arbitrarily choose to have two layers of 10 neurons and a dropout of 0.2 for each.

```
def create_mlp_2layer():
    model_arch = [
    Dense(10, activation="relu"),
    Dropout(0.2),
    Dense(10, activation="relu"),
    Dropout(0.2),
    Dense(1)
    ]
    model = Sequential(model_arch)
    model.compile(optimizer="adam", loss ='mse')
    return model
```

CS 1.6 Step 6 – Running Our Model

Now we can create a model, we should run it and compare its performance against our testing set.

```
model = create_mlp_2layer()
early_stopping = EarlyStopping(monitor="loss",
patience=3)
model.fit(X_train, y_train, batch_size=10, epochs=20,
callbacks=[early_stopping])
test_mse = model.evaluate(X_test, y_test)
rmse = np.sqrt(test_mse)
```

Using this model you should be able to get an root mean squared error (RMSE) of around 1.12. This is not great, so how might we improve it? One way might be to normalise the target values, another might be to improve our architecture or simply run for more epochs. We will let you experiment with different modelling methods (including normalisation, early stopping, etc.) and focus on putting what we have learned about Bayesian optimisation to use to decide on two variables – the dropout rate and the number of neurons per layer.

CS 1.7 Step 7 – Automatically Finding an Optimised Architecture Using Bayesian Optimisation

In the final chapter of the book, we looked at how we could use Bayesian optimisation to optimise our models, and now is a chance to put that knowledge to good use.

```
def improvement(y_pred, y_best, sigma):
    imp = (y_best - y_pred) / sigma
    return imp

def exp_imp(y_pred, y_best, sigma):
    imp = improvement(y_pred, y_best, sigma)
    e_i = exp_imp = sigma * (imp * norm.cdf(imp) +
norm.pdf(imp))
    return e_i
```

We need to update the previous construction function to be a function of neurons and dropout

```
def create_mlp_2layer(n1, n2, d1, d2):
    model_arch = [
    #keras.layers.Dense(512, activation="relu"),
    Dense(n1, activation="relu"),
    Dropout(d1),
    Dense(n2, activation="relu"),
    Dropout(d2),
    Dense(1)
    ]
    model = Sequential(model_arch)
    model.compile(optimizer="adam", loss ='mse')
    return model
```

We can now compile this into an evaluation function which returns the RMSE as a function of the architectural parameters.

```
def eval_model(params):
    n1, n2, d1, d2  = params
    model = create_mlp_2layer(n1, n2, d1, d2)
    early_stopping = EarlyStopping(monitor="loss",
patience=3)
    model.fit(X_train, y_train, batch_size=10,
epochs=20, callbacks=[early_stopping])
```

```
# check our RMS error on the data to get an idea of
our model's performance
    test_mse = model.evaluate(X_test, y_test)
    rmse = np.sqrt(test_mse)
    return rmse
```

Let us now look at constructing a Gaussian process in scikit-learn

```
kernel = RBF() + WhiteKernel(noise_level=1e-4)
gp = gpr(kernel=kernel)
```

We can create a domain which covers all the possible combinations of neurons and dropout:

```
n1 = np.arange(1,100,1)
n2 = np.arange(1,100,1)
d1 = np.arange(0,1,0.1)
d2 = np.arange(0,1,0.1)

domain = [[a,b,c,d] for a in n1 for b in n2 for c in d1
for d in d2]
```
Let's randomly sample that to initialize the search:
```
idx = np.random.choice(np.arange(len(domain)), size=(3))

X_init = [domain[i] for i in idx]
y_init = [eval_model(x) for x in X_init]
y_init = np.array(y_init).reshape(-1,1)
```

We of course need to remove any points we initialised with from the pool:

```
domain = np.delete(domain, idx, axis=0)
```

Now, let us combine everything into a search for 10 epochs (you can see how changing this number changes the search).

```
num_epochs = 10
for e in range(num_epochs):
    gp.fit(X_init, y_init)
    y_best = np.min(y_init)
    y_pred, sigma = gp.predict(domain,
return_std=True)
    acquisition = exp_imp(y_pred, y_best, sigma)
    idx = np.argmax(exp_imp)
    X_new = domain[idx]
```

```
y_new = eval_model(X_new)
X_init = np.vstack((X_init, X_new))
y_init = np.vstack((y_init, y_new))
domain = np.delete(domain, idx, axis=0)
```

You can of course use this technique to search over more parameters using what you learned in the final chapter of the book.

Case Study 2

Time Series Forecasting with LSTMs

In this case study, we will use explore how two different kinds of long short term memory (LSTM) architectures can be used for time series forecasting. We will be using the Sunspots dataset, and developing models to predict sunspots over time. This dataset has been released into public domain, and can be obtained from Kaggle here: https://www.kaggle.com/robervalt/sunspots/data

CS 2.1   Simple LSTM

The first model we will explore is a simple LSTM. We first load and configure our data, before defining our model architecture with Keras. We will also choose how long we want our input and output sequences to be, here using input sequences of length 24 – corresponding to 24 months of data – and output sequences of 6 (six months of data). In this case we 'are using our LSTM to directly generate floating-point output sequences. As such, mean squared error (MSE) is a sensible loss function. We again use the Adam optimiser. While we can use MSE (or root mean squared error (RMSE)) to get an impression of how well our network is doing, here we will use the "plot_-data" function. This function allows us to easily get an impression of how well our model is doing by overlaying our predicted sequences onto our ground-truth sequences. Additionally, this allows us to see how well the model does as we increase the time interval for the predictions – from one month up to n months. Let us take a look at the code:

Deep Learning for Physical Scientists: Accelerating Research with Machine Learning,
First Edition. Edward O. Pyzer-Knapp and Matthew Benatan.
© 2022 John Wiley & Sons Ltd. Published 2022 by John Wiley & Sons Ltd.

```python
import numpy as np
import matplotlib.pyplot as plt
import tensorflow as tf
from tensorflow import keras
from sklearn.preprocessing import StandardScaler

"""
Simple example of using LSTM for time series
forecasting.
"""

def load_data(train_prop=0.8, normalise=False):
    """
    Load data from the 'sunspots.csv'. This can be
obtained from Kaggle:
    https://www.kaggle.com/robervalt/sunspots/data

    train_prop: the proportion of data in our training
set, default 0.8
    """
    data = np.loadtxt("sunspots.csv", delimiter=",",
skiprows=1, usecols=(2))
    train_end = int(data.shape[0]*train_prop)
    data_train = np.reshape(data[0:train_end], (-1, 1))
    data_test = np.reshape(data[train_end::], (-1, 1))
    scaler = StandardScaler()
    if normalise:
        data_train = scaler.fit_transform(data_train)
        data_test = scaler.transform(data_test)
    n_classes = max(data.reshape(-1))+1
    return data_train, data_test, n_classes

def config_data(data, i_len, o_len):
    """
    Configure data
    data: time series data
    i_len: input sequence length
```

```
    o_len: output sequence length
    """
    t_len = i_len + o_len
    X = []
    y = []
    data = data[0:int(data.shape[0]/(t_len))*t_len]
    for i in range(data.shape[0]-t_len):
        X.append(data[i:i+i_len])
        y.append(data[i+i_len:i+t_len])
    X = np.array(X).reshape(-1, i_len, 1)
    y = np.array(y).reshape(-1, o_len, 1)
    return X, y

def plot_data(y_test, y_pred, plot_steps=None):
    """
    Plot data to visualise how our model is doing.
    y_test: ground-truth data
    y_pred: predicted data
    plot_steps: which steps to plot for the sequence -
allows easy visualisation
                of how our network is doing at predicting
further into the
                future.
    """
    n_plots = y_test.shape[1]
    if plot_steps == None:
        plot_steps = list(range(n_plots+1))
    for i in range(n_plots):
        if (i+1 in plot_steps):
            plt.subplot(len(plot_steps), 1, plot_steps.
index(i+1)+1)
            plt.plot(y_test[:,i].reshape((-1, 1)))
            plt.plot(y_pred[:,i].reshape((-1, 1)))
            plt.ylabel("sunspots")
            plt.title("offset: " + str(i+1) + " months")
    plt.xlabel("time")
    plt.show()
```

```
# define length of input sequences and length of
output sequences
i_len = 24
o_len = 6

# load and configure data from sunspots dataset
data_train, data_test, n_classes = load_data()
X_train, y_train = config_data(data_train, i_len,
o_len)
X_test, y_test = config_data(data_test, i_len, o_len)

# define our network architecture
model_arch = [
keras.layers.LSTM(64),
keras.layers.Dropout(0.1),
keras.layers.Dense(o_len)
]
model = keras.Sequential(model_arch)

# compile and train our network
model.compile(loss="mean_squared_error",
              optimizer=tf.optimizers.Adam(),
              metrics=["mean_squared_error"])

model.fit(X_train, y_train,
          batch_size=32,
          epochs=100,
          validation_split=0.2)

y_pred = model.predict(X_test)

# plot our values, looking at how well we predict for
the shortest longest time
# intervals
plot_data(y_test, y_pred, [1, o_len])
```

Running this code should produce the following plot:

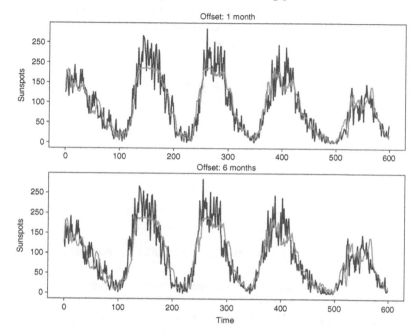

Here, we can see that our model does fairly well at predicting the general trend, but loses some of the high frequency information, resulting in a "smoothed" version of the sequence. This smoothing is common in forecasting models, as it is often difficult to learn some of the more specific, high-frequency components of the data. There are various strategies we can employ to improve the performance of our model. These include increasing the number of training epochs, introducing normalisation, modifying our loss function and/or optimiser, and trying another architecture. Why not modify the code to see if some of these strategies could help improve performance? In the next section, we will investigate how a more sophisticated sequence-to-sequence (or seq2seq) model can be used for the same problem.

CS 2.2 Sequence-to-Sequence LSTM

In this example we will see how a seq2seq model can be used for time series forecasting. This example will build on our previous case study – we will be using the same data, with the same goal: to predict time series sunspot data. This time, however, we will be using a seq2seq model, comprising an encoder and decoder, and we will be explicitly writing the code for the autoregression process to generate sequential output, rather than the simple multi-output approach used in the previous example. This example will also demonstrate how to use one-hot-encoded data in a time-series forecasting setting. Let us take a look at the code:

```python
import numpy as np
import matplotlib.pyplot as plt
import tensorflow as tf
from tensorflow import keras
from sklearn.preprocessing import StandardScaler

"""
Example of seq2seq modelling for time series
forecasting.
"""

def load_data(train_prop=0.8, normalise=False):
    """
    Load data from the 'sunspots.csv'. This can be
obtained from Kaggle:
    https://www.kaggle.com/robervalt/sunspots/data

    train_prop: the proportion of data in our training
set, default 0.8
    """
    data = np.loadtxt("sunspots.csv", delimiter=",",
skiprows=1, usecols=(2))
    data = np.int64(data)
    train_end = int(data.shape[0]*train_prop)
    data_train = np.reshape(data[0:train_end], (-1, 1))
    data_test = np.reshape(data[train_end::], (-1, 1))
    scaler = StandardScaler()
    if normalise:
        data_train = scaler.fit_transform(data_train)
        data_test = scaler.transform(data_test)
    n_classes = max(data.reshape(-1))+1
    return data_train, data_test, n_classes

def config_data(data, i_len, o_len, n_classes):
    """
    Configure data
    data: time series data
```

```
    i_len: input sequence length
    o_len: output sequence length
    """
    t_len = i_len + o_len
    X = []
    y = []
    data = data[0:int(data.shape[0]/(t_len))*t_len]
    for i in range(data.shape[0]-t_len):
        X.append(data[i:i+i_len])
        y.append(data[i+i_len:i+t_len])
    X = np.array(X).reshape(-1, i_len, 1)
    y = np.array(y).reshape(-1, o_len, 1)
    X = tf.keras.utils.to_categorical(X, n_classes)
    y = tf.keras.utils.to_categorical(y, n_classes)
    X_dec = np.zeros_like(y)
    X_dec[:,1::] = y[:,0:-1]
    return X, X_dec, y


def plot_data(y_test, y_pred, plot_steps=None):
    """
    Plot data to visualise how our model is doing.
    y_test: ground-truth data
    y_pred: predicted data
    plot_steps: which steps to plot for the sequence -
allows easy visualisation
                of how our network is doing at predicting
further into the
                future.
    """
    n_plots = y_test.shape[1]
    if plot_steps == None:
        plot_steps = list(range(n_plots+1))
    for i in range(n_plots):
        if (i+1 in plot_steps):
            plt.subplot(len(plot_steps), 1, plot_steps.
index(i+1)+1)
            plt.plot(y_test[:,i].reshape((-1, 1)))
            plt.plot(y_pred[:,i].reshape((-1, 1)))
            plt.ylabel("sunspots")
            plt.title("offset: " + str(i+1) + " months")
    plt.xlabel("time")
    plt.show()
```

```python
def get_models(classes_in, classes_out, n_hidden):
    """
    Define & return our encoder, decoder, and combined
Keras model.
    classes_in: number of unique tokens in input
    classes_out: number of unique tokens in output
    n_hidden: size of hidden layer
    """
    # define encoder part of network
    encoder_inputs = keras.layers.Input(shape=(None,
classes_in))
    encoder = keras.layers.LSTM(n_hidden,
return_state=True)
    encoder_outputs, state_h, state_c =
encoder(encoder_inputs)
    encoder_states = [state_h, state_c]
    encoder_model = keras.models.Model(encoder_inputs,
encoder_states)

    # define decoder part of network
    decoder_inputs = keras.layers.Input(shape=(None,
classes_out))
    decoder_lstm = keras.layers.LSTM(n_hidden,
return_sequences=True, return_state=True)
    decoder_outputs, _, _ = decoder_lstm
(decoder_inputs, initial_state=encoder_states)
    decoder_dense = keras.layers.Dense(classes_out,
activation='softmax')
    decoder_outputs = decoder_dense(decoder_outputs)

    # define our model with both encoder and decoder
inputs
    model = keras.models.Model([encoder_inputs,
decoder_inputs], decoder_outputs)

    # define decoder data flow
    decoder_state_input_h = keras.layers.Input(shape=
(n_hidden,))
    decoder_state_input_c = keras.layers.Input(shape=
(n_hidden,))
    decoder_states_inputs = [decoder_state_input_h,
```

```
decoder_state_input_c]
    decoder_outputs, state_h, state_c = decoder_lstm
(decoder_inputs, initial_state=decoder_states_inputs)
    decoder_states = [state_h, state_c]
    decoder_outputs = decoder_dense(decoder_outputs)
    decoder_model = keras.models.Model
([decoder_inputs] + decoder_states_inputs,
[decoder_outputs] + decoder_states)

    return model, encoder_model, decoder_model


def predict_sequence(encoder_model, decoder_model, X,
n_steps, n_classes):
    """
    Function to predict sequences using encoder &
decoder models
    encoder_model: encoder model (see above)
    decoder_model: decoder model (see above)
    X: input data
    n_steps: number of steps to predict in output
    n_classes: number of unique tokens in input
    """
    output = []
    for x in X:
        x = x.reshape((1, -1, n_classes))
        state = encoder_model.predict(x)
        X_dec = np.zeros_like(x).reshape(1, -1,
n_classes)
        y_pred_all = []
        for t in range(n_steps):
            y_pred, h, c = decoder_model.predict
([X_dec] + state)
            y_pred_all.append(y_pred[0,0,:])
            state = [h, c]
            target_seq = y_pred
        y_pred_all = np.array(y_pred_all)
        output.append(y_pred_all)
    return np.array(output)

def decode_one_hot(encoded):
    """
```

```
    Simple function for decoding our one-hot encoded
data
    """

return np.array([[np.argmax(x) for x in X] for X in
encoded])

# define length of input sequences and length of
output sequences
i_len = 24
o_len = 6

# load data from sunspots dataset
data_train, data_test, n_classes = load_data()

# configure data and get seq2seq model components
X_train, X_train_dec, y_train = config_data
(data_train, i_len, o_len, n_classes)
X_test, X_test_dec, y_test = config_data(data_test,
i_len, o_len, n_classes)
model, encoder_model, decoder_model = get_models
(n_classes, n_classes, 64)

# compile model
model.compile(loss="categorical_crossentropy",
              optimizer=tf.optimizers.Adam(),
              metrics=["accuracy"])

# train our model
model.fit([X_train, X_train_dec], y_train,
          batch_size=32,
          epochs=500,
          validation_split=0.2)

# predict our output values
y_pred = predict_sequence(encoder_model,
decoder_model, X_test, o_len, n_classes)

# decode and plot our values, looking at how well we
predict for the shortest
# longest time intervals
y_pred = decode_one_hot(y_pred)
```

```
y_test = decode_one_hot(y_test)
plot_data(y_test, y_pred, [1, o_len])
```

Here, we can see a few key changes from our simple LSTM example. Firstly, our data configuration now converts our input and output data into one-hot encoded data, and generates a decoder input that we use for training the decoder. Due to the seq2seq model requiring two inputs for training (to train the encoder and decoder), we cannot use a simple Keras Sequential() object to instantiate our network. As such, our architecture definition is more sophisticated, and provides a handle on our encoder, decoder, and our combined model which we will use for training. Another key point here is that, as we are now using one-hot encoded data, we need to adapt our loss function accordingly: now using the categorical cross-entropy to evaluate our network loss. Our predictions are no longer quite as straightforward either – we now use our predict_sequence() function to autoregressively predict our output given our input; iteratively updating our output for each step, and using this as the input for the next step. Given the more sophisticated nature of the network, and the more complex embedding of our data, this network will take significantly more time to train than the previous example. Try running the code and taking a look at the output – you should have something similar to the following:

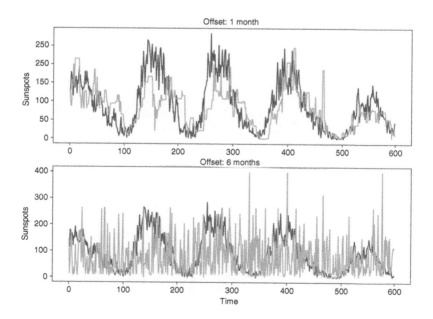

As we can see here, our seq2seq model clearly performs worse than our original model. There are a number of factors contributing to this, however the most significant factor is the increase in model complexity: while we have increased the complexity of our model (and the representation of our data) we have not increased the number of training epochs. Thus, it is not surprising that the model performs more poorly. Furthermore, while seq2seq models are essential in certain applications, sometimes a simpler model is perfectly sufficient!

Case Study 3

Deep Embeddings for Auto-Encoder-Based Featurisation

In this case study, we will explore how autoencoders can be used to create deep embeddings – a kind of featurisation achieved through exploiting the latent representation of a neural network, in this case an auto-encoder. Just as with the auto-encoder example, we will train the network to reproduce its input:

```
from tensorflow.keras import datasets, layers, models
import tensorflow as tf
from sklearn.ensemble import RandomForestClassifier
from sklearn.model_selection import train_test_split
import numpy as np

# We use RDKit to load molecular representations and
to convert these to
# adjacency matrices
from rdkit.Chem import MolFromSmiles
from rdkit.Chem import AllChem as ac

datafile = "nr-er.smiles"
f = open(datafile, "r")
data = f.readlines()
f.close()

feats = []
X = []
y = []
```

Deep Learning for Physical Scientists: Accelerating Research with Machine Learning,
First Edition. Edward O. Pyzer-Knapp and Matthew Benatan.
© 2022 John Wiley & Sons Ltd. Published 2022 by John Wiley & Sons Ltd.

```
# Get adjacency matrix from each SMILES molecule and
the label
# indicating whether it's toxic (1) or non-toxic (0)

# This code also fixes the size of our input matrices
to 132 x 132
# as graphs of different sizes will result in variable
sizes of
# adjacency matrices.

for line in data:
    try:
        splitline = line.split("\t")
        smiles = splitline[0]
        mol = MolFromSmiles(smiles)
        A = ac.GetAdjacencyMatrix(mol)
        x = np.zeros((132, 132))
        x[:A.shape[0], :A.shape[1]] = A
        X.append(x)
        y.append(int(splitline[-1]))
    except:
        pass
X = np.array(X)
X = X.reshape((X.shape[0], X.shape[1], X.shape[2], 1))

# Split into training and validation data
X_train, X_test, y_train, y_test = train_test_split
(X, y, test_size=0.20)

# Create model class for VAE, inheriting from tf.
keras.Model
class VAE(tf.keras.Model):
    def __init__(self, latent_dims):
        super(VAE, self).__init__()
        self.latent_dims = latent_dims
        # Create encoder part of the network
        self.encoder = models.Sequential()
        self.encoder.add(layers.InputLayer
(input_shape=(132,132,1)))
        self.encoder.add(layers.Conv2D(filters=64,
kernel_size=3, strides=(2, 2), activation='relu'))
        self.encoder.add(layers.Conv2D(filters=32,
```

```
kernel_size=3, strides=(2, 2), activation='relu'))
        self.encoder.add(layers.Flatten())
        self.encoder.add(layers.Dense(latent_dims*2))

        # Create decoder part of the network
        self.decoder = models.Sequential()
        self.decoder.add(layers.InputLayer
(input_shape=(latent_dims,)))
        self.decoder.add(layers.Dense(units=33*33*32,
activation=tf.nn.relu))
        self.decoder.add(layers.Reshape(target_shape=
(33,33,32)))
        self.decoder.add(layers.Conv2DTranspose
(filters=32, kernel_size=3, strides=2,
activation='relu', padding='SAME'))
        self.decoder.add(layers.Conv2DTranspose
(filters=64, kernel_size=3, strides=2,
activation='relu', padding='SAME'))
        self.decoder.add(layers.Conv2DTranspose
(filters=1, kernel_size=3, strides=1,
activation='relu', padding='SAME'))

    def gen_sample(self, decode_inti=None):
        if decode_init == None:
            decode_init = tf.random.normal(shape=
(latent_dims*2))
        return self.run_decoder(decode_init,
apply_sig=True)

    def reparam(self, mean, logv):
        rsample = tf.random.normal(shape=mean.shape)
        return rsample * tf.exp(logv * 0.5) + mean

    def run_encoder(self, x):
        mean, logv = tf.split(self.encoder(x),
num_or_size_splits=2, axis=1)
        return mean, logv

    def run_decoder(self, z, apply_sig=False):
        logits = self.decoder(z)
        if apply_sig:
            return tf.sigmoid(logits)
        return logits
```

```python
    def log_normal_pdf(self, sample, mean, logv,
raxis=1):
        return tf.reduce_sum(-0.5 * ((sample-mean) **
2.0 * tf.exp(-logv) + logv + (tf.math.log(2. * np.
pi))))


    def loss(self, x):
        mean, logv = self.run_encoder(x)
        z = self.reparam(mean, logv)
        x_logits = self.run_decoder(z)
        cross_entropy = tf.nn.
sigmoid_cross_entropy_with_logits(logits=x_logits,
labels=np.float32(x))
        logpx_z = -tf.reduce_sum(cross_entropy, axis=
[1,2,3])
        logpz = self.log_normal_pdf(z, 0.0, 0.0)
        logqz_x = self.log_normal_pdf(z, mean, logv)
        return -tf.reduce_mean(logpx_z + logpz -
logqz_x)


    def update_grads(self, x, optimizer=tf.keras.
optimizers.Adam()):
        with tf.GradientTape() as tape:
            loss = self.loss(x)
        grads = tape.gradient(loss, self.
trainable_variables)
        optimizer.apply_gradients(zip(grads, self.
trainable_variables))


    def train(self, X_train, epochs, batch_size=10):
        for epoch in range(epochs):
            i = 0
            while i < X_train.shape[0]-batch_size:
                x = X_train[i:i+batch_size].reshape
(batch_size, 132, 132, 1)
                self.update_grads(x)
                i += batch_size


epochs = 100
latent_dim = 100
n_examples = 10
```

```
model = VAE(latent_dim)
model.train(X_train, epochs)
```

Unlike in the original example, in this case we are not looking to generate new data using the network – instead, we want to use the network to transform our features using the latent space:

```
# run the encoder on X - we only care about the mean
values in this case
X_train_feats, _ = model.run_encoder(X_train)
```

In our original variational auto-encoders (VAE) example, we did not use our y data at all – as we were only interested in creating a generative model to produce new molecular structures. In this case, we want to use our *y* values to train a model to predict drug toxicity. Using neural networks for feature representation in this way is often an effective method for reducing dimensionality – meaning we can then use models that may not have coped well otherwise. In this case, we are able to exploit the two-dimensional structure of our data through the use of convolutional neural networks (CNNs) in order to intelligently reduce dimensionality. We can now train a model to predict toxicity using our new features:

```
# train a new model on our features

tox_model = RandomForestClassifier() tox_model.fit
(X_train_feats, y_train)
accuracy = model.score(X_test, y_test)
```

Now that we know how to use deep embeddings from neural networks, there are a couple of other things you may want to consider:

1) What happens when you change the parameters of the neural network – can this help to improve the representation, and thus the performance of the toxicity prediction model? Experiment by changing a few simple parameters first, starting with the size of the latent space (latent_dim).

2) In this case, we used convolutional layers to exploit the spatial information in our data. Other applications would benefit from different types of neural network layers – what do you think may work well with a text classification problem?

Index

Note: Page numbers in *italic* refer to figures, page numbers in **bold** refer to tables.

Deep Learning for Physical Scientists: Accelerating Research with Machine Learning,
First Edition. Edward O. Pyzer-Knapp and Matthew Benatan.
© 2022 John Wiley & Sons Ltd. Published 2022 by John Wiley & Sons Ltd.